FREE Study Skills DVD Offer

Dear Customer,

Thank you for your purchase from Mometrix! We consider it an honor and privilege that you have purchased our product and want to ensure your satisfaction.

As a way of showing our appreciation and to help us better serve you, we have developed a Study Skills DVD that we would like to give you for FREE. **This DVD covers our "best practices" for studying for your exam, from how to use our study materials to how to prepare for the day of the test.**

All that we ask is that you email us your feedback that would describe your experience so far with our product that we can post on our website in order to help other customers make a sound buying decision. Good, bad or indifferent, we want to know what you think!

To get your **FREE Study Skills DVD**, email freedvd@mometrix.com with "FREE STUDY SKILLS DVD" in the subject line and the following information in the body of the email:

a. The name of the product you purchased.

b. Your feedback. It can be long, short, or anything in-between, just your impressions and experience so far with our product. A good testimonial will include how our study material met your needs and will highlight features of the product that you found helpful.

c. Your name as you would like for it to be displayed with the testimonial.

d. Your full name and shipping address where you would like us to send your free DVD, along with your email address and phone number (for shipping purposes only).

If you have any questions or concerns, please don't hesitate to contact me directly.

Thanks again!

Sincerely,

Jay Willis
Vice President
jay.willis@mometrix.com
1-800-673-8175

TABLE OF CONTENTS

Top 20 Test Taking Tips 4
Foundations of Statistics and Sampling Distributions 5
Probability 35
Inferential Statistics and Correlation and Regression 63
Practice Test 105
  Practice Questions 105
  Answers and Explanations 119
Secret Key #1 - Time is Your Greatest Enemy 136
  Pace Yourself 136
Secret Key #2 - Guessing is not Guesswork 136
  Monkeys Take the Test 136
  $5 Challenge 137
Secret Key #3 - Practice Smarter, Not Harder 138
  Success Strategy 138
Secret Key #4 - Prepare, Don't Procrastinate 138
Secret Key #5 - Test Yourself 139
General Strategies 139
Special Report: What Your Test Score Will Tell You About Your IQ 145
Special Report: What is Test Anxiety and How to Overcome It? 147
  Lack of Preparation 147
  Physical Signals 148
  Nervousness 148
  Study Steps 150
  Helpful Techniques 151
Special Report: Retaking the Test: What Are Your Chances at Improving Your Score? 156
Special Report: Additional Bonus Material 158

Top 20 Test Taking Tips

1. Carefully follow all the test registration procedures
2. Know the test directions, duration, topics, question types, how many questions
3. Setup a flexible study schedule at least 3-4 weeks before test day
4. Study during the time of day you are most alert, relaxed, and stress free
5. Maximize your learning style; visual learner use visual study aids, auditory learner use auditory study aids
6. Focus on your weakest knowledge base
7. Find a study partner to review with and help clarify questions
8. Practice, practice, practice
9. Get a good night's sleep; don't try to cram the night before the test
10. Eat a well balanced meal
11. Know the exact physical location of the testing site; drive the route to the site prior to test day
12. Bring a set of ear plugs; the testing center could be noisy
13. Wear comfortable, loose fitting, layered clothing to the testing center; prepare for it to be either cold or hot during the test
14. Bring at least 2 current forms of ID to the testing center
15. Arrive to the test early; be prepared to wait and be patient
16. Eliminate the obviously wrong answer choices, then guess the first remaining choice
17. Pace yourself; don't rush, but keep working and move on if you get stuck
18. Maintain a positive attitude even if the test is going poorly
19. Keep your first answer unless you are positive it is wrong
20. Check your work, don't make a careless mistake

Foundations of Statistics and Sampling Distributions

Survey vs.experiment

In a survey, a surveyor collects responses from a sample or population. In a good survey, the surveyor does not intervene with responses. There are often no treatments to any group, or controlled variables. However, a survey can be used prior to and/or following an intervention. A survey can be utilized to examine the effects of an experiment.

In an experiment, the experimenter controls certain variables on treatment and non-treatment groups and observes the results, making comparisons between the groups. There can be one or more controlled variables and one or more uncontrolled observations that result. The goal of the experimenter is to identify correlations, models of good fit, and cause-effect relationships.

Census

A census is a survey that collects information from every member of a population. It is, therefore, implicit that such a population must be finite. The United States Census is conducted every ten years as required by the U.S. Constitution in order to determine the number of congressional representatives assigned to each district of each state. In principal, every member of the population must be included in this census, however, that is not practically possible, so, in addition to the Census (mailers to each home), an independent Census Coverage Survey is conducted (in-person interviews by field staffers). The results from both surveys are used to generate missing data from people not covered by either survey.

Sample survey

A sample survey consists of obtaining data from a subset of the population. The data are a measure of a variable. The data gathered may be discrete or continuous or categorical. An advantage of sampling over a census is that the entire population does not need to be included in the study.

To obtain the most accurate measure of the mean requires sampling without bias, since bias will shift the estimated value of the mean.

When the variable being measured has different means and/or variance for its subsets, a more accurate estimate of the mean and the variance of the population can be obtained by a method called stratification, wherein the population is divided into more homogeneous subsets and the sample selected contains the same proportion of each subset as contained in the population.

The number of data points in the sample must only be large enough to obtain estimates of the parameters within pre-determined confidence intervals.

Conducting an experiment

The general purpose of an experiment is to determine the causal relationship, if any, between variables. Generally, this is accomplished by an experiment wherein the experimenter controls one or more "controlled variables" and then records one or more "observations".

The simplest form of an experiment involves bivariate data (X, Y), where X is the controlled variable and Y is the observation. Generally, a function is "fitted" to the data so that it predicts the value of Y when given the value of X. The function is of the form, $\hat{y}=f(x)$, where $\hat{y}$ is the predicted value of the observation, Y, for a given value x of the controlled variable, X. For example, suppose $\hat{y} = f(x) = 3 + 4x$. In one run of the experiment, $(x_2, y_2) = (2{,}10.5)$, so $\hat{y} = 3 + 4(2) = 11$ and the residual is "observed value" – "predicted value" $= 10.5 - 11.0 = -0.5$.

Controlled variables and observations

The experimenter controls one or more "controlled variables" and then records one or more "observations". The simplest case is bivariate data (X, Y), where X is the controlled variable and Y is the observation.

For example, suppose the following equation is fitted to the data, $\hat{y} = f(x) = 3 + 4x$, where $\hat{y}$ is the value predicted for a given value of x, $\hat{y}$ is called the dependent variable because for every value of the independent variable x, $f(x)$ assigns one and only one value of $\hat{y}$.

Assume the experimenter chooses Y as the controlled variable and X as the observation, which would result in a fitted equation of the form, $\hat{x} = g(y)$. However, if Y is "braking distance" and X is "reaction time", then, when the experimenter attempts to control "braking distance", he is forced to control "reaction time". Therefore, the experimenter determines that X is the true "controlled variable", there is a causal relationship, and the equation is arranged so that X is the independent variable and Y is the dependent variable ($\hat{y} = 3 + 4x$).

Controlled and observational studies

A controlled study, includes one or more treatment and control groups and attempts to examine the effects of one or more interventions on an outcome. The main emphasis of a controlled study is the implementation of an intervention.

An observational study may or may not be a part of a controlled study. An observational study can examine random participant groups or intact groups (quasi-experimental). Observational studies focus on observation of different groups; this observation may occur after some intervention and thus serve as part of a controlled study, or it may occur separate from any type of experiment and/or randomization.

Characteristics of a well-conducted survey

The characteristics of a well-designed and well-conducted survey are the selection of sample units or clusters with probabilistic methods that can provide estimates of sampling error, and the absence of bias.

A unit/participant is an element of the population (i.e., one person).

A probabilistic method is consists of constructing a sample frame with a selection procedure and a data collection method. A sample frame may consist of the entire population or an identifiable subset of the population (i.e., names in a telephone directory, names from the DMV, social security numbers, electoral register). The sampling error is expressed as a confidence interval or a margin of error. For example, a 95% confidence interval for a mean says there is a 2.5% chance the mean is above the interval and a 2.5% chance it is below the interval. The margin of error is half the width of the confidence interval.

Major sources of bias are under-coverage bias, non-response bias, selection bias, and measurement bias.

Note: A population in a survey is generally not homogeneous with respect to the observations being made; rather, it often consists of homogeneous subsets.

Various types of surveys

Different surveys vary in timing, sampling, mode of administration, and mode of data collection. The timing can be cross-sectional (data collected from sample units once), longitudinal (data collected from sample units at different intervals whose pattern is identical for each unit), and time-series (data collected from sample units at different intervals whose pattern may be different for each unit). The sampling may be simple random, stratified, systematic or clustered. In addition, sampling may be multi-stage, in which the sampling technique is repeated on increasingly smaller groups down to the sample unit. The mode of administration may be a researcher administered survey or a self-administered survey. The mode of data collection maybe telephone, mail, online, in-home, or mall intercept.

Sample size, selection procedure, sample frame, and data collection method

A good set of sample units is randomly selected, and it is of appropriate sample size to achieve the desired margin of error (margin of error = 1/2 the width of the confidence interval).

There are several types of selection procedures: simple random sampling, stratified sampling, cluster sampling.

A sample frame contains an identifiable subset of the population (i.e., names in a telephone directory, names from the DMV, social security numbers, electoral register).

The data collection method refers to the agreed upon rules of collecting data from the sample units. It includes such things as asking a question in precisely the same way each time and without bias, noting non-responses, etc.

Simple random sampling

A simple random sampling of n sample units is a sampling done in such a way that any subset of the population containing n sample units has the same probability of being selected. This means that the sample frame must be the entire population itself (otherwise some subsets of size n could not be selected), and the population must be relatively homogeneous.

Simple random sampling may be done with or without replacement. Random variables must be independent and identically distributed (iid). Identically distributed simply means each comes from the same population. Independent means that selection of one variable cannot depend on whether other variables have already been selected, therefore, sampling without replacement means the sample units are not independently selected. However, if the population is large enough this is not a problem, since the probability of being selected twice is negligible.

Process of stratification

When subsets of a population vary considerably, each subset should be sampled. The process of stratification groups members of the population into relatively homogeneous subsets or strata. The sample frame for stratification is the entire population – no element of the population can be excluded. Once stratification is completed, then random or systematic sampling is performed on each stratum. There are two types of stratified sampling strategies: Proportionate allocation is done by randomly selecting a number of sample units from each stratum that is in proportion to the number of units it contains. This strategy would result in a more representative set of statistics for the observation(s). Optimum allocation is done by randomly selecting a number of sample units from each stratum that is in proportion to the size of its variance. This strategy would result in a more accurate set of statistics for the observation(s), since it would minimize the overall variance. Both of the above methods are forms of stratified random sampling.

Systematic sampling and sample frame

Systematic sampling is performed on a sample frame that consists of the entire population in which each unit is arranged in order. Examples of order: the street number and block number of a population consisting of all houses in a city, the order in which customers enter a store. Systematic sampling is then conducted on the sample frame by selecting sample units at equally spaced intervals of the ordered sequence of units. Example of systematic sampling on a sample frame: every tenth customer entering a store is selected. Systematic sampling should only be used when the population is relatively homogeneous.

Cluster sampling

The union of all clusters should be the entire population. This is written as, $\bigcup_{i=1}^{i=N} A_i = P$ where P is the entire population divided into N subsets, A_i.

The intersection of any two clusters must be empty. This is written as,
$A_i \cap A_j = 0$ for any $i \neq j$.

Thus, the sample frame is the entire population.

The differences in definition between stratified sampling and cluster sampling are: (i) each cluster must be as heterogeneous as possible in order to represent the entire population, whereas each stratum should be relatively homogeneous, (ii) random sampling is used to select the clusters from which data will be collected (viz., selected clusters are themselves sample units), whereas random sampling is used to select the sample units in each stratum.

The difference in objective between stratified sampling and cluster sampling is the former strives for precision by identifying homogeneous subgroups, while the latter strives to save costs by not requiring information unique to each subgroup.

Control

Control over the sample size of homogeneous subsets of a population can be achieved through stratification and proportional sampling. This will help ensure that each subset is adequately represented by setting the number of sample units proportional to the size of the subset.

Generally, control is better for a researcher-administered survey than for a self-administered survey, because a researcher has greater control over the environment in which the data is collected. Control over response rate is worse for data collection by mail or online. Control over response rate is better for data collection by personal in-home surveys, because a skilled interviewer can increase the response rate.

Bias

Bias can be defined as experimental bias arising out of the selection of sample units in a survey (selection bias), the measurement of the responses (systematic bias), the response of the sample units (response bias and non-response bias), and experimenter's bias which can affect the selection of sample units, the responses and the data collection.

Bias can also be defined as estimator bias. If an equation $\hat{X}$ estimates a variable X, then $\hat{X}$ is a bias estimator if $E(\hat{X}) \neq E(X)$. Simply stated, if the mean of the estimation is not equal to the true amount, it is biased.

Example:
$\hat{\mu} = \frac{\sum_{i=1}^{i=n} x_i}{n}$ is an unbiased estimator of μ,

because $E(\hat{\mu}) = E\left(\frac{\sum_{i=1}^{i=n} x_i}{n}\right) = \mu = E(X)$.

Example:

$\hat{\sigma}^2 = \frac{\sum_{i=1}^{i=n} x_i^2 - n\bar{x}^2}{n}$ is a biased estimator of σ^2,

because $E(\hat{\sigma}^2) = E\left(\frac{\sum_{i=1}^{i=n} x_i^2 - n\bar{x}^2}{n}\right) = \frac{n-1}{n}\sigma^2 \neq \sigma^2 = E(\sigma^2)$.

Therefore, $\frac{\sum_{i=1}^{i=n} x_i^2 - n\bar{x}^2}{n-1}$ is an unbiased estimator of σ^2.

Selection Bias occurs when there are errors in selecting sample units. Examples are self selection, elimination of sample units that dropped out of the study, and under-coverage bias where one or more subsets of the population are not included in the survey.

Systematic bias occurs when the method of measurement is biased (i.e., incorrectly calibrated thermometer).

Response Bias occurs when people tend to give a particular response which is not according to their beliefs or the facts.

Non-Response Bias occurs when non-respondents, if they had responded, would have tended to have a different response than respondents.

Experimenter's Bias occurs when the experimenter directly manipulates the sampling, the responses or the data. Examples are pre-screening sample units and the loaded question.

Experimental and sampling unit

An experimental unit is the baseline piece of a study, i.e., a person, group, class, region, etc. Some researchers attend to the idea that an experimental unit is the larger set, with a sampling unit representing a subset of the experimental unit. Other researchers equate the two phrases and use them interchangeably to denote the simplest group being studied, with the sampling unit, simply constituting a sample of that unit, gleaned from the population.

Treatment group, control groups, and positive and negative controls

A treatment group is a randomly selected group of experimental units that receive treatments. A treatment is a controlled variable, also known as a factor. There may be one or more levels of a factor.

A control group is the group that does not receive a treatment.

A positive control group is one that represents a positive effect of the experiment, when comparing the two groups. For example, if you predict that a new mathematical software will significantly increase student learning, the group using only math textbooks serves as a positive control group, illustrating the positive effect of the experiment or intervention.

A negative control group is one that represents a negative effect of the experiment on the two groups. A control group is not always needed. For example, if a controlled variable is the turbine speed to measure the observed current and voltage outputs, it would not make sense to have a control group with zero turbine speed, since the current and voltage outputs are already known to both be zero.

Randomization

The types of randomization used in an experiment are simple random sampling, cluster sampling, and stratified sampling. Simple random sampling is the most common sampling technique; it simply means that all samples of the same number of experimental units have an equal probability of being selected from the population. You may assign a number to each individual in a population. Randomly choose a number, using a random number generator, number on a chart, etc; those individuals with that number will be in the sample. Simple random sampling will randomly distribute any unknown variable contained in the experimental units. This will still not guarantee an exactly uniform distribution of the unknown variable among the treatment and control groups, but as the number of experimental units increases, a uniform distribution will be approached. Cluster sampling divides a homogenous population into clusters or groups. Stratified sampling divides the population into groups that are each homogenous in nature; groups are chosen based on some a priori characteristic.

Confounding variable

A confounding variable is a variable that is not part of an experiment and cannot be controlled. Examples of confounding variables are age, gender, years teaching, etc. Such a variable can act as a covariate in a study, therefore serving as a predictor variable.

Eliminating confounding

Confounding occurs when an explanatory variable can be expressed as a function of another confounding variable. This confounding variable may or may not have a cause-effect relationship with the observations, however, the experimenter does not wish to consider it in the experiment. To eliminate the effect of this confounding variable, the experimenter could assign experimental units with the confounding variable to both treatment and control groups. The experimenter could also eliminate the effect of this confounding variable by only selecting experimental units that possess this confounding variable. For example, if the experimenter desired to determine if smoking increased heart disease, but also knew that age was a confounding variable that also influenced the incidence of heart disease, then a cohort study would only contain subjects (cohorts) of a similar age.

Placebo effect, blinding and double blinding

The placebo effect occurs when the subject's perception and expectation affects the outcome. (Placebos were commonly administered by Hyde Park physicians to their wealthy patients in London in the 19th century.)

The effect can be positive if the person expects a treatment as helpful, and negative if the expectation is harmful. The placebo effect can be separated from the treatment effect by using a blind placebo control.

Blinding is accomplished when the human experimental units do not know which group they are in – a treatment group or a placebo group. This eliminates the placebo effect.

Double blinding is accomplished when the human subjects do not know their group assignments, and the experimenter does not know the group assignments of the human subjects until after the observations are recorded.

Replication

Replication is the multiple repetition of a treatment or control for the purpose of estimating the variance. Replication is not a repetition of the measurement of an observation from the same experimental unit, rather it requires the observation from another experimental unit.

The three major sources of bias in experiments is little or no randomization in the selection of experimental units, the placebo effect (which can only occur with cognitive subjects, i.e. humans in clinical trials), and confounding.

Experimental units may contain unknown or uncontrollable variables that could affect the observations in one direction. These effects will tend to be cancelled out as the number of experimental units is increased and simple random selection is used.

The placebo effect can affect the outcome in one direction if the human subjects expect the treatment to be helpful or harmful. It is eliminated by implementing a blind study wherein the subjects are not informed of their group assignments.

An explanatory variable is a function of one or more controlled variables, but it may also be a function of a confounding variable which the experimenter does not wish to consider.

Biological, pharmaceutical or clinical experiments

Due to biological variance, many biological, pharmaceutical and clinical experiments require separate treatment groups and a control group. Specifically, this is due to the fact that organisms can possess different forms at a genetic locus (alleles) producing different forms of enzymes or blood types from individual to individual. [This can even lead to seeming outliers, which are, in fact, valid data points, and cannot be discarded.]

A treatment group contains experimental units that are identical to those in a control group. The experimental units in a treatment group are subjected to one or more treatments that constitute the controlled variable(s).

A control group contains experimental units that are not subjected to the treatments. Rather, they may be treated with a placebo which, in a blind study, is not known to them, and in a double blind study, it is not known to the experimenter(s), as well.

Control groups are needed to help control the effects of biological variance (that, and a large sample to increase the likelihood of selecting experimental units with rare biological variance, i.e., Phase III in drug development).

Total variance, treatment variance and error variance of an experiment

The total variance is the sum of the squared deviations divided by the degrees of freedom (n minus 1)

The variance of a treatment is the sum of squared deviations divided by the degrees of freedom (n minus 1; also called error variance).

Error variance is the percentage of variance that is not accounted for in a study; i.e., the total variance minus the variance of each of the treatments and the control. In other words.

Note: If X, Y and Z are independent variables, $Var(X + Y + Z) = Var(X) + Var(Y) + Var(Z)$ and

$$Var(Total) = Var(Treatment\ 1) + Var(Treatment\ 2) + Var(Error)$$

Example of an experiment that is randomized, has a control group, has two treatments (including the control) and has replicates for each treatment

Example: Insecticide F is applied to half the plots of soil. The other half receives no fertilizer and is the Control C. F and C are assigned to plots randomly. There are 4 replicates of each of the two treatments (F and C). The difference of the average of the two treatments is 15.8. The total variance of all the data is computed. The variance of just the C data and just the F is computed. $Var(Total) = Var(C) + Var(F) + Var(Error)$, and

$$Std\ Error\ of\ Difference = \sqrt{\frac{Var(Error)}{\#\ of\ Replicates}} = \pm 3.2.$$

PLOTS ARE NUMBERED		RESULTS GROUPED BY PLOT	
1	2	F,9	C,18
3	4	F,4	C,24
5	6	C,30	F,16
7	8	C,29	F,9

RESULTS GROUPED BY TREATMENT	C	F
	18	9
	24	4
	30	16
	29	9
TREATMENT AVERAGE	25.3	9.5
GRAND AVERAGE	17.4	

TOTAL VARIANCE	94.3
VARIANCE WITHIN CONTROL	30.3
VARIANCE WITHIN F	24.3
VARIANCE OF ERROR	39.7
DIFFERENCE BETWEEN CONTROL AND F	15.8
STD ERROR OF DIFFERENCE BETWEEN CONTROL AND F	± 3.2

Example of a randomized matched pairs design, and indicate why it can be better than a two treatment design without matching

Four of twenty plots of soil are randomly selected. Corn is planted in each of the four plots and the average height measured after one month (I, Before Treatment). More corn is planted with fertilization and the average height measured after one month (A, After Treatment). The average difference of the matched pairs is 15.8.

$$Var(\bar{A} - \bar{I}) = Var(A) + Var(\bar{I}) - 2Cov(\bar{A}, \bar{I})$$
$$= \frac{Var(A)}{n} + \frac{Var(I)}{n} - 2\sqrt{Var(A)Var(\bar{I})}Corr(A_i, I_i)$$

Where n is the number of matched pairs (4) and $Corr(A_i, I_i) = \frac{\sum_{i=1}^{i=4}(A_i-\bar{A})(I_i-\bar{I})}{(n-1)\sqrt{Var(A)Var(I)}}$.

The purpose of a matched pairs design over a two treatment design is to reduce the variance of the difference in the average before and after values when the correlation is positive. In fact, using the same data as in a two treatment design without matching, we have reduced the std. dev. of the difference in the averages from ±3.2 to ±1.5. A two-treatment design does not account for differences in the outcome, caused as a result of similarity in the groups. However, a matched pairs design does account for these similarities, i.e., husband – wife; pre-test/post-test, etc.

BEFORE TREATMENT	AFTER TREATMENT
I,9	A,18
I,4	A,24
I,16	A,30
I,9	A,29

	RESULTS GROUPED BY TREATMENT	
	I	A
	9	18
	4	24
	16	30
	9	29
TREATMENT AVERAGE	9.5	25.3
GRAND AVERAGE	17.4	

TOTAL VARIANCE	94.3
VARIANCE WITHIN I	24.3
VARIANCE WITHIN A	30.3
$\bar{A} - \bar{I}$	15.8
CORRELATION (A,I)	0.21
VARIANCE OF ($\bar{A} - \bar{I}$)	2.4
STD. DEV. ($\bar{A} - \bar{I}$)	± 1.5

Creating a randomized block design

Simple random sampling is used to select experimental units for the control group and the treatment group, after which the treatment group is divided into more homogeneous blocks. Blocking of experimental units is done to reduce variance within the blocks. The experimenter divides the experimental units into blocks that are more homogeneous with respect to a characteristic that the experimenter assumes will mean that the observations within each block will have less variance than the variance computed for all the observations.

The block design is described below, where block 1 is the control and blocks 2-4 each contain experimental units whose increase in homogeneity over the population as a whole will result in less variance within each block.

$y_i = \alpha_0 + \alpha_1 Z_{1i} + \alpha_2 Z_{2i} + \alpha_3 Z_{3i} + \alpha_4 Z_{4i} + \varepsilon_i$where
y_i is the outcome for the ith unit,
α_0 is a constant (the grand average of all the observations)
α_1 is the mean difference between the treatment and the control
α_{2-4} are the effects of being in block 2, 3, or 4
ε_i is the error, what has to be added to make the right side of the equation the same as the left.

Observational studies, surveys, and experiments

Observational studies, is sometimes seen as studies that lack random selection of units and control groups and may only be considered useful as starting points for a controlled study with random selection of units. Other researchers deem observational studies as those that may indeed use randomly selected groups, but choose to observe the groups, whether with or without an intervention.

surveys and experiments are controlled studies in the sense that the units are often randomly selected and all attempts are made to reduce or eliminate bias, with the exception being quasi-experimental groups

Surveys and experiments can look at effect changes; surveys can provide quantitative and qualitative data, in experimental and non-experimental situations.

Populations of both surveys and experiments may be homogeneous or may consist of homogeneous subsets.

Statistical quantification for both surveys and experiments includes estimating means, variance and confidence intervals for the whole population, including subsets.

Stratification of units in a survey and blocking of units in an experiment both strive to create more homogeneous subsets with reduced variance.

Exploratory data analysis

Exploratory data analysis (EDA) seeks to first develop hypotheses about patterns (central tendency, clusters, gaps, modality) observed in univariate data (using dot plots, stem-and-leaf plots, histograms, box plots), or cause and effect in bivariate data (using scatter plots).

Second, assumptions are assessed in determining the appropriate choice for the underlying distribution (i.e., normal, binomial, etc.), if any.

Third, a survey (opinions or facts collected from a population) or experiment (a controlled study with dependent and independent variables, comparison of treatments, random allocation of units to treatments, replication, blocking, blinding, etc.) is designed to collect data from which statistical inference will be used (from estimators, confidence intervals, hypothesis testing, etc.) to draw conclusions and/or propose another survey or experiment.

Quantitative and categorical variables

A quantitative variable is numerical and may be discrete or continuous. A discrete variable can take on countably many values. That is, there is a one-to-one correspondence between each discrete variable and all the positive integers or some subset of them. For example, $X = \{1,4,5,7\}$, means that X can have the values 1, 4, 5, and 7. A continuous variable must be able to assume any value in a specified interval of the set of real numbers. For example, $X \in [0,1]$, means the variable X can take on any value from 0 to 1; similarly, $X \in [0, \infty)$, means X can take on any value from 0 to infinity.

A categorical variable is a name or code representing an object. For example, "red" and "other" are categorical variables representing red sports cars and all other cars, respectively. One can use a bar graph to compare the average speed of red sports cars and all other cars in one week end along a stretch of highway. Categorical variables are also known as nominal variables (presumably because the categories can be identified by names).

Discrete quantitative variable

A discrete variable can only have countably many possible values on a given interval of the set of real numbers. A countable set (of numbers) is a set with the same cardinality as some subset of the set of natural numbers (all the positive integers {1, 2, 3, 4, 5, 6, 7, 8...}). In other words, one must be able to assign a natural number (in sequence, starting with 1) to each outcome.

An example of a discrete variable with a finite number of outcomes is the experiment in which a die is tossed. There are six possible outcomes for the variable x, so the cardinality of the outcomes is 6, $\{x_1, x_{2,} x_3, x_{4,} x_5, x_6\} = \{1,2,3,4,5,6\}$.

An example of a discrete variable with an infinite number of outcomes is the experiment in which a coin is repeatedly tossed until the tail shows up. The possible outcomes for the variable x are, $\{x_1, x_{2,} x_3, x_{4,} x_5, x_6, \ldots\} = \{T, HT, HHT, HHHT, HHHHT, HHHHHT, \ldots\}$

It is possible (though not likely) that one possible outcome is that the head shows up a countably infinite number of tosses in a row. The cardinality of the outcomes is the same as the set of all natural numbers. We say the cardinality is countably infinite.

Continuous quantitative variable

A continuous quantitative variable may assume any value on a given interval of the set of real numbers.

An example is a machine that stamps out metal rods whose length is never less than 20 mm and never more than 23 mm. The continuous variable is the length, and it may assume any value on the interval [20, 23]. This may also be written $0 \leq x \leq 23$. (Note that the probability density function has not been defined. For example, it could be continuous uniform or normal.)

One may also determine if a variable is continuous by just looking at its cumulative distribution function (CDF). A variable is continuous if its cumulative distribution function (CDF) is continuous.

Example: The probability of selecting any number such that $1 \leq x \leq 6$ has the following

$$\text{CDF, } F(x) = \begin{matrix} 0 \text{ for } x < 1 \\ \frac{x-1}{5} \text{ for } 1 \leq x \leq 6 \\ 1 \text{ for } x \geq 6 \end{matrix}$$

The CDF is continuous because there is not a break in the function when graphed; it is a continuous graph.

Univariate and bivariate data

Univariate data has one number, or variable, for each datum or data point.

Example: A sample of trees is cut, and the number of tree rings is recorded for each. This is an example of discrete univariate data.

Example: One tree is cut and the width of each tree ring is recorded. This is an example of continuous univariate data.

Bivariate data has two variables for each data point.

Example: A sample of trees is cut, and the number of tree rings and the known age in years is recorded for each. This is an example of discrete bivariate data. A scatter plot would show a strong positive linear correlation in the data. In fact, a well-designed experimental study would reveal that there is a causal relationship between each pair of bivariate data – the number of years of growth causes the creation of an equal number of tree rings.

Example: The average income of populations of different ethnic groups is displayed using a bar graph. The bivariate data pairs a categorical variable (ethnicity) with a continuous quantitative variable (income).

Dot plot and stem plot

A dot plot graphs the individual univariate data points along the x-axis. The dot plot can display outliers and patterns such as clusters and gaps. The dots lie vertically above their value on the x-axis. The size of the dot is analogous to the width of the bars in a histogram. In other words, since the diameter of the dots projects an interval onto the x-axis, the larger the dots, the more points may be included in each vertical array of dots.

Therefore, the dot size can affect the shape of the dot plot. If the dot size is too great patterns may be obscured, and if the dot size is to small spurious patterns may emerge. The optimum dot size can be determined by minimizing the error between the values predicted by the dot plot and the underlying data; however, this requires advanced software. Generally, the smaller the dot size, the greater is the amount of data required to support the conclusions regarding patterns in the shape of the dot plot.
A stem plot and discuss the effect of the stem on patterns in the shape.

A stem plot is a data representation that presents numbers in two parts. First the univariate data is arranged in order of size (i.e., 132, 145, 230, 235, 440, 442, 452, 819). Next, one or two of the last digits of each datum are assigned to a leaf (i.e., 32,45,30,35,40,42, 52, 19), and the remaining digits are assigned to the stem (i.e., 1,1,2,2,4,4,4,8). Using our example,

1 | 32 45
2 | 30 35
3 |
4 | 40 42 52
5 |
6 |
7 |
8 | 19

Key: 2 | 30 = 230, Leaf Unit: x1, Stem Unit: x100

The numbers on the stem are arranged in order of increasing size (from top to bottom). The numbers on the leaves are arranged in order of increasing size (from left to right).

The "Key" indicates how the data is parsed between the stems and leaves; the "Leaf Unit" indicates the factor by which each number on the leaves must be multiplied; the "Stem Unit" indicates the factor by which each number on the stems must be multiplied.
The stem plot can display outliers and patterns such as clusters and gaps. In our example, there is a possible outlier on stem 8 (datum is 819) and a gap on stem 3.

Histogram

A histogram is a graphical method in which quantitative univariate data is placed in bins or bars whose width represents the range of values for the data placed into the bin/bar and whose height is proportional to the number of data points contained therein.

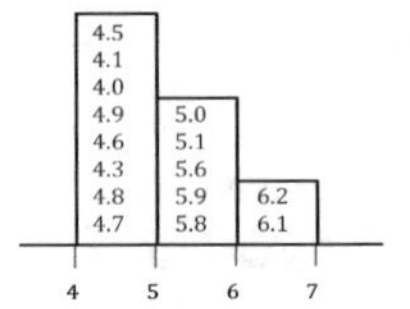

The histogram can display outliers and patterns such as clusters and gaps. The width of the bar can affect the shape of the histogram. If the width is too great, patterns may be obscured, and if the width is too small, spurious patterns may emerge.

Bar width can be determined by examining the range of the data and deciding on an appropriate number of intervals. The interval size can then be adjusted to fit the desired number of intervals; 10 intervals are often considered a reasonable number. The optimum bar width can be determined by minimizing the error between the values predicted by the histogram and the underlying data; however, this requires advanced software. Generally, the smaller the bar width, the greater is the amount of data required to support the conclusions regarding patterns in the shape of the histogram.

Bar graph

A bar graph is constructed by placing bars of length proportional to the quantity of the data for each data group either horizontally or vertically. The data is typically discrete (shoe sizes) or categorical (shoe color).

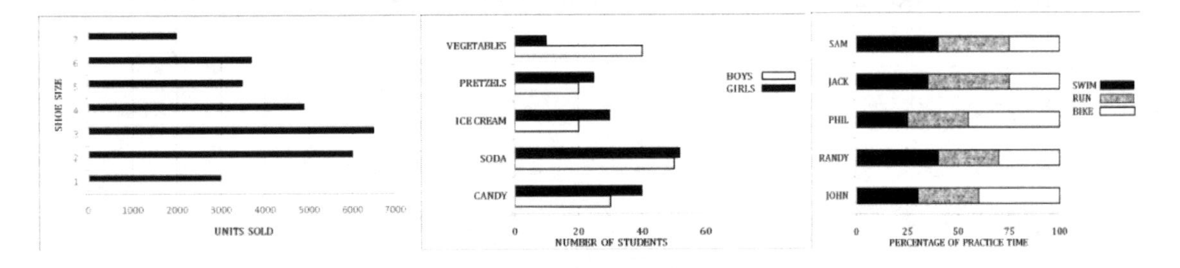

Two or more bars may be grouped together to compare subsets of the categories (i.e., the categories are five different types of snack foods eaten by students, the subsets are male and female students, and the quantity being measured for each category is the number of male or female students who consume each category of snack food).

Bar graphs may be stacked to show subsets of the quantity of each category (i.e., the categories are the names of different contestants in a triathlon, the quantity being measured is the amount of practice time devoted by each contestant, the subset of this quantity is the amount of practice time devoted to each of the three events in the triathlon).

Cumulative frequency, cumulative percentage, and cumulative distribution

The cumulative frequency of a variable is simply the reported sum of all previous frequencies. The cumulative percentage of a variable is the product of 100 and the ratio of the current cumulative frequency to the total cumulative frequency; the final listed cumulative frequency should be 100%. The cumulative distribution of a variable X=A is the fraction of all the data for which X<A. In the above example, the cumulative distribution, F_C, for X=30 is 10/20 or 0.50.

Frequency

The frequency of a variable X=A is simply the number of times X=A appears in the data set. A histogram of a set of data is a type of frequency plot, wherein the height of each bar is proportional to the number of times data appears in the set with a value in the interval which is proportional to the width of each bar. For example, the width of a bar spans the interval of the variable 5 to 10 and there are 20 data points that fall within this interval, the height of the bar is graphed proportional to 20. In addition to histograms that plot the frequency of data that fall in various intervals, histograms can also plot the fraction of data that fall in various intervals. In this case the sum of each interval times the height of the bar divided by the range is 1. This is written as:

$$\sum_{i=1}^{i=N} \frac{\text{width of } ith \text{ interval}}{\text{range}} \times (\text{height of bar}) = 1.$$

(Recall that the range is the difference between the maximum and minimum datum.)

Mode, modality and clusters

The mode is the most frequently occurring value in a univariate data set. Informally, the mode is the peak. For example, in the univariate data set {1,2,2,5,7,9,11}, the median is 5 (it divides the data in half) and the mode is 2 (the most frequently occurring datum). The modality refers to the number of pronounced peaks shown in a histogram. If a histogram reveals the appearance of more than one pronounced peak, it is bimodal.

Clusters may be found in a scatter plot of bivariate data. For example, a marketing company will use cluster analysis to separate the general population of consumers into market segments or clusters so that each cluster will have low variance with respect to the observed data.

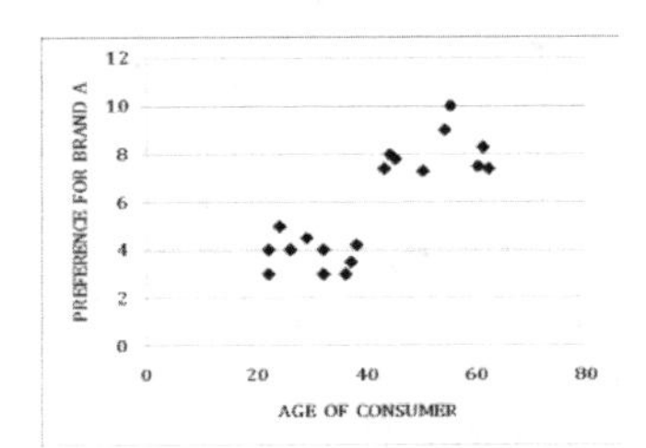

(Note: cluster analysis is not to be confused with cluster sampling, where each cluster is supposed to be representative of the general population, meaning each cluster must be chosen so that its variance is high enough to be close to that of the general population.)

Scales of measure

An ordinal scale of measurement uses the order of the numbers as a way to measure. For example, a patient is asked to rate his pain on a scale of 0 to 5, where 0 is no pain and 5 is unbearable pain. Other examples include movie, hotel and restaurant ratings using stars – the more stars, the better.

An interval scale of measurement measures a quantity in intervals with an arbitrary zero. For example, the Centigrade scale has intervals where each degree is the increase in temperature when one Calorie is added to one gram of water at one atmosphere, but the zero point is arbitrarily set at the freezing point of water.

The ratio scale of measurement measures a quantity in intervals with zero set where the measure is truly zero. For example, the mass of a substance is only zero when there is no substance. Another example, the Kelvin temperature scale measures the temperature in intervals and has a zero point defined as that point where the atoms have no motion (alternatively, when a volume of gas has zero pressure).

Central tendency

The common measures of central tendency are the mean, the median and the mode. These measures of central tendency are not necessarily equal in value, but each has unique properties that make it useful.

The mode is simply the most frequently occurring value, informally referred to as the peak in the data.

The mean, typically the arithmetic average, minimizes the expected value of the squared deviation. This is written as,
$E[(X - X_{mean})^2] \leq E[(X - A)^2]$, for any value A of the variable X.

The median is the value for which half the data is smaller and half the data is larger, and it minimizes the expected value of the absolute deviation. This is written as,
$E[|X - X_{median}|] \leq E[|(X - A)|]$, for any value A of the variable X.

Dispersion

The common measures of dispersion are variance, standard deviation, range, and interquartile range (IQR).

The variance has the general expression,
$Var(X) \equiv E\left[\left(X - E(X)\right)^2\right]$, where E(X) is the expected value of the variable X, and the variance is the expected value of the square of the difference between X and E(X). For a normal distribution, E(X) is the average or mean. The common notation for variance is σ^2. The common notation for the estimate of the variance is s^2. The variance simply relates the variability in scores, as related to the mean.

The standard deviation is the square root of the variance. The common notation for the standard deviation is σ. The common notation for the estimate of the standard deviation is s.

The range is the difference between the highest and lowest data point.

The interquartile range is Q3-Q1. The interquartile range reports the difference between the median of the lower 50% and the median of the upper 50%, i.e., the difference between the 25th and 75th percentiles.

Shape

The common measures of shape are skewness and kurtosis.

Skewness is defined as $\frac{E[(X-\mu)^3]}{\sigma^3}$. It will contribute asymmetry to the shape of the plotted data because 3 is an odd number, causing data points equidistant from μ to have different contributions to the shape. When the skewness is positive, the right tail is longer because the expectation is that $(X - \mu)$ will be positive. Conversely, when skewness is negative, the left tail is longer.

Kurtosis is defined as $\frac{E[(X-\mu)^4]}{\sigma^4}$. It will NOT contribute asymmetry to the shape of the plotted data because 4 is an even number, causing data points equidistant from μ to have identical contributions to the shape. When the kurtosis is positive, the peak is more acute and higher than a normal distribution. When the kurtosis is negative, the peak is lower and wider than a normal distribution.

Position

The common measures of position are q-quantiles, quartiles, percentiles and z-scores.

For data in order of increasing size, the kth q-quantile is the smallest data point for which the fraction of data points smaller than or equal to it is k/q (where k and q must be integers, q is fixed and k=1,2...q).

When q=4, the q-quantile is called a quartile. Measures of position for the quartile are the 1st quartile (Q1), the 2nd quartile (Q2), and the 3rd quartile (Q3). Twenty five percent of the data is below Q1, 50% of the data is below Q2, and 75% of the data is below Q3.

When q=100, the q-quantile is called a percentile. Measure of position for the percentile is any integer percent from 1 to 100. For example, 10% of the data is below the 10th percentile, and 90% of the data is below the 90th percentile.

The z-score of a variable x is $z = \frac{x-\mu}{\sigma}$. The terms μ and σ are the population average and standard deviation (square root of the variance, σ^2) of a normal population. For example, if $\mu = 10$ and $\sigma = 2$, the z-score for $x = 9$ is

$$z = \frac{x - \mu}{\sigma} = \frac{9 - 10}{2} = -0.5$$

Interval and ratio scale measures

The following measures of a distribution cannot be done with variables using the interval scale of measurement; rather, they must be done with the ratio scale of measurement.

These measures are:

Measures of Centrality: geometric mean, harmonic mean

Measure of Dispersion: coefficient of variance, z-score

Measures of Shape: skewness, kurtosis

The essential difference between interval scales and ratio scales is that the interval scale has an arbitrary zero while the ratio scale has zero corresponding to a true zero of the quantity being measured.

Some common scales of measurement that are ratio scales are: mass, length, volume, velocity, acceleration, and Kelvin scale of temperature.

Some common scales of measurement that are interval scales are: Centigrade or Fahrenheit scales of temperature, IQ tests, and the year date on a calendar.

Examples: Ratios of temperatures in the Centigrade scale have no meaning (because the zero point is arbitrary), although ratios of intervals do have meaning. Ratios of temperatures in the Kelvin scale do have meaning (because the zero corresponds to a true measure of no temperature when the atoms cease all movement).

Shape parameter and skewness

The shape of a probability density function (pdf, or distribution) is characterized by the skewness and the kurtosis, which are both defined in terms of the central moments.

The general equation for the central moments is,

$\mu_k = E[(X - \mu)^k]$, $k = 2,3,4$ where μ_k is the kth moment about the mean, and μ is $E[X]$, the expectation value for X for any pdf for which E[X] is defined.

Note that
$\mu_2 = E[(X - E[X])^2]$ is the general definition for the variance, σ^2, of any distribution for which E[X] is defined.
The skewness is defined as $\gamma_1 = \frac{\mu_3}{\sigma^3}$.

It may be estimated from a sample of n data points, $\gamma_1 \approx \frac{n^2}{(n-1)(n-2)} \frac{\mu_3}{\sigma^3}$.

Note that skewness will contribute asymmetry to the shape of the plotted data because k is an odd number, meaning that data points equidistant from μ will have different contributions to the shape. When the skewness is positive, the right tail is longer because the expectation is that $(X - \mu)$ will be positive. Conversely, when skewness is negative, the left tail is longer.

Kurtosis

The kurtosis is defined as, $\gamma_2 = \frac{\mu_4}{\sigma^4}$. It may be estimated from a sample of n data points,

$$\gamma_2 \approx \frac{n^2 - 1}{(n-1)(n-3)} \left[\frac{\mu_4}{\sigma^4} - 3 + \frac{6}{n+1} \right].$$

Note that kurtosis will NOT contribute asymmetry to the shape of the plotted data because k is an even number, meaning that data points equidistant from μ will have identical contributions to the shape.

When the kurtosis is positive, the peak is more acute and higher than a normal distribution, and the tails are higher than a normal distribution. An example is the Laplace distribution,

$$L(x, \mu, \beta) = \frac{1}{2\beta} e^{\left(-\frac{|x-\mu|}{\beta}\right)}, \text{mean} = \mu = 5, \text{variance} = 2\beta^2 = \frac{10}{4}, \text{skewness} = 0, \text{kurtosis} = 3.$$

The Laplace distribution is overlaid with the normal distribution in the graph below. Both distributions have mean=5, variance=10/4.

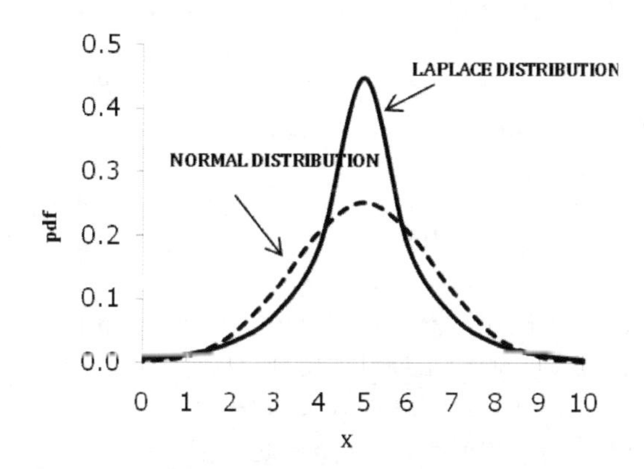

When the kurtosis is negative, the peak is lower and wider than a normal distribution, and the tails are lower than a normal distribution. An example is the Continuous Uniform distribution,

$$f(x) = \frac{1}{b-a} \text{ for } a \leq x \leq b, \text{ and } f(x) = 0 \text{ for } x < a \text{ or } x > b.$$

$$\text{mean} = \frac{a+b}{2}, \text{ variance} = \frac{1}{12}(b-a)^2, \text{ skewness} = 0, \text{ kurtosis} = -\frac{6}{5}$$

where the parameters, a and b, are adjusted to 2.2614 and 7.7386, resp., so that the mean and variance are the same as those for the normal distribution in the plot below. The Continuous Uniform distribution is overlaid with the normal distribution. Both distributions have mean = 5 and variance = 10/4.

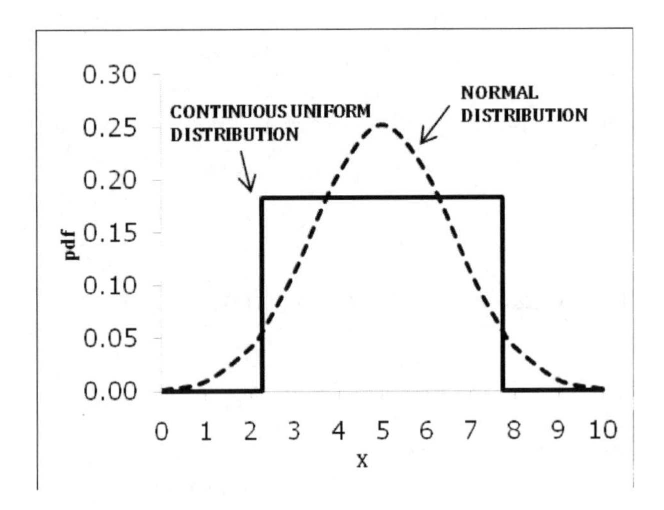

Symmetry

Symmetry is a characteristic of the shape of the plotted data. Specifically, it refers to the operation of taking the mirror image along a vertical axis through the median.

Skewness will contribute asymmetry to the shape of the plotted data because k is an odd number, meaning that the data points equidistant from μ will not have identical contributions to the shape.

Kurtosis will not contribute asymmetry to the shape of the plotted data because k is an even number, meaning that the data points equidistant from μ will have identical contributions to the shape.

When the kurtosis is zero (as are any higher moments), the distribution is symmetrical about the average.

Outlier

An outlier is an extremely high or extremely low value in the data set. It may be the result of measurement error, in which case, the outlier is not a valid member of the data set. However, it may also be a valid member of the distribution. Unless a measurement error is identified, the experimenter cannot know for certain if an outlier is or is not a member of the distribution. There are arbitrary methods that can be employed to designate an extreme value as an outlier. One method designates an outlier (or possible outlier) to be any value less than $Q_1 - 1.5(IQR)$ or any value greater than $Q_3 + 1.5(IQR)$.

Biological data, in particular, may produce seeming outliers due to biological diversity. For example, in a study measuring the rate of drug metabolism in rats, a potential outlier from one rat may well be a valid data point if the rat belongs to a small subset of rats with a markedly different metabolism (i.e., a missing or altered enzyme).

Summarizing the central tendency, dispersion, shape and position of distributions of univariate data

The measures used to summarize central tendency are mean, median and mode.

The measures used to summarize the dispersion are variance, standard deviation, range, and interquartile range.

The measures used to summarize the shape are skewness and kurtosis.
The measures used to summarize the position are quantiles and z–scores.

Some of these measures, like the mean, variance, standard deviation, z-scores, skewness and kurtosis, can be estimated with equations assuming that the data comes from a population with a normal distribution. The equations for these measures will generally be different if the population does not have a normal distribution (i.e., continuous uniform, Poisson, Gamma).

Other measures, like the quartiles, percentiles, IQR, cumulative frequency and median can be evaluated without any assumption of the underlying distribution function, and, in fact, there may be no simple mathematical function which simulates the population (i.e. the particle size distribution of air quality samples with multiple peaks).

Mean and median of a set of univariate data

The arithmetic mean, $\bar{x}$, is defined as the sum of all the data points divided by the number of data points, $\bar{x} = \sum_{i=1}^{i=n} x_i$, where i is the ith point and n is the number of data points. The median may be loosely defined as the value for which half the data is above it and half the data is below it. The more precise definition is as follows. For an odd number of data, the median is the datum that is in the middle of the data set when it is arranged in order of size, $x_1, x_2, \ldots x_{i-1}, x_i, x_{i+1}, \ldots x_n$, where n is odd, $x_{i+1} \geq x_i$ for $1 \leq i \leq n-1$, and $(i-1) = (n-i)$,
$i = \frac{(n+1)}{2}$ so that $x_{\frac{(n+1)}{2}}$ is the median.

Example: For data set $\{5,2,7,1,8,3,7\}$, we rearrange is order of increasing size to $\{1,2,3,5,7,7,8\}$; $n = 7$, and $x_{\frac{(n+1)}{2}} = x_4 = 5$ is the median.

For an even number of data, the median is the mean of the two middle data points when the data set is arranged in order of size, $x_1, x_2, \ldots x_{i-1}, x_i, \ldots x_n$, where n is even, $x_{i+1} \geq$

x_i for $1 \leq i \leq n-1$, and the number of data points on either side of the two middle points is $\frac{n-2}{2}$, and $\frac{n-2}{2}+1=(i-1), i=(n+2)/2$ so that $\frac{(x_{n/2}+x_{(n+2)/2})}{2}$ is the median.

Example: For data set= $\{5,2,7,1,8,3,7,10\}$ we rearrange to $\{1,2,3,5,7,7,8,10\}$; $n=8$, and $\frac{(x_{n/2}+x_{(n+2)/2})}{2}=\frac{(x_4+x_5)}{2}=\frac{(5+7)}{2}=6$ is the median.

Variance and its estimate

The variance has the general expression, $Var(X) \equiv E\left[\left(X-E(X)\right)^2\right]$, where E(X) is the expected value of the variable X, and the variance is the expectation value of the square of the difference between X and E(X). For a normal distribution, E(X) is the average or mean.

The normal distribution is a mathematical function of x, and it contains two constants or parameters, the average (μ) and the variance (σ^2), $N(x,\mu,\sigma^2)=\frac{1}{\sqrt{2\pi\sigma^2}}e^{-\frac{(x-\mu)^2}{2\sigma^2}}$. The parameters μ and σ^2 are the precise values for the mean and variance, respectively. However, even though a sample of data comes from a normally distributed population, these values are not generally known. The estimate for the average, $\bar{x}$, and the estimate for the variance, s^2, depends on the sample set and the number of data points, $\bar{x}=\frac{1}{n}\sum_{i=1}^{i=n}x_i$, $s^2=\frac{1}{n-1}\sum_{i=1}^{i=n}(x_i-\bar{x})^2$. The standard deviation, σ, is the square root of the variance. The estimate of the standard deviation is $s=\sqrt{\frac{1}{n-1}\sum_{i=1}^{i=n}(x_i-)^2}$.

As a consequence of the median dividing the upper 50% of the data from the lower 50% of the data, the median minimizing the expectation value of what expression?

The expected value of $|X-c|$ is minimized when c is the median, x_{median}.

This can be written, $E[|X-x_{median}|] \leq E[|X-c|] \leq$where c can be any possible value for x.

Example: For data set $\{1,2,3,5,7,7,8\}$; $n=7$, the median is 5 and the average is 4.714. $\frac{1}{6}\sum_{i=1}^{i=7}\frac{|x_i-x_{median}|}{6} \cong 2.327$, and $\sum_{i=1}^{i=7}\frac{|x_i-x_{average}|}{6} \cong 2.714$

However, the expected value of $(x-c)^2$ is minimized when c is the average, $x_{average}$. Example: For the same data set, $E[(X-x_{median})^2] \cong 7.667$ and $E\left[\left(X-x_{average}\right)^2\right] \cong$ 7.571

Recall that the variance is the expected value of the square of the deviation of the variable X from the average.

Quantile

Given a set of data placed in order of increasing size. The kth q-quantile is the smallest data point for which the fraction of data points smaller than or equal to it is k/q (where k and q must be integers, q is fixed and k=1,2...q).

For example, given the data set arranged in order of increasing size {1,3,4,5,8,9,10,11,14,18}, we want to divide the data into 4 quantiles, so q=4 and k=1,2,3,4. [Note: a 4-quantile is referred to as a quartile.] Next, we want the value of the first quartile. In this case, k/q = ¼, and ¼ of the data consists of 10/4=2.5 data points. In order to get at least ¼ of the data, we must round up to 3 data points. That will include data points 1,3,4. Thus, the value 4 is the first quartile, since ¼ of the data is smaller or equal to it.

Similarly, the second quartile is 8, the third quartile is 11 and the fourth quartile is 18.

When q=100, the quantile is referred to as a percentile. Percentiles are often used with standardized testing. A score of the 99th percentile relates that 99% of the scores were lower than that particular score. A birth weight in the 17th percentile relates that 17% of birth weights are lower than this baby's birth weight.

Interquartile range

A box plot of a quantitative variable will display any outliers, the lowest non-outlier, the first quartile (Q_1), the second quartile (Q_2), the third quartile (Q_3), and the highest non-outlier.

The first quartile is that value for which 25% of the data (not including outliers) is smaller.

The second quartile is the median (not including outliers); therefore, 50% of the data is below the second quartile.

The third quartile is that value for which 25% of the data (not including outliers) is larger.

The box plot is also known as also known as a box and whisker display.

The interquartile range (IQR) is $Q_3 - Q_1$.
Example: The first quartile is 10, the second quartile is 20, and the third quartile is 35. The interquartile range is $35 - 10 = 25$. Twenty five per cent of the data is below 10, 50% of the data is below 20, and 75% of the data is below 35.

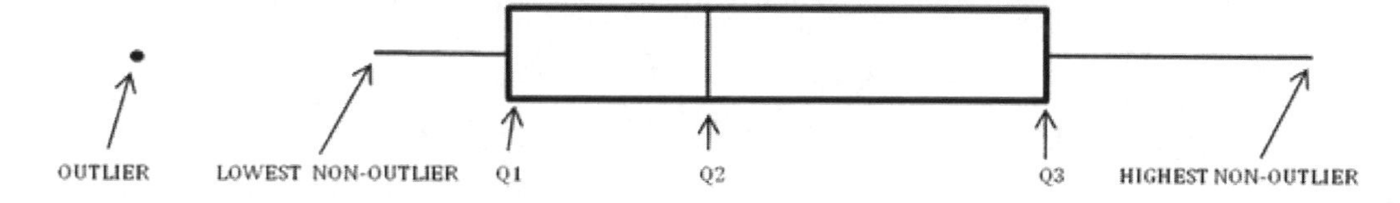

Standard score or z-score

The z-score of a variable x is $z = \frac{x-\mu}{\sigma}$. The terms μ and σ are the population average and standard deviation (square root of the variance, σ^2). They are not estimates calculated from a sample of the population. They can only be calculated from the entire population (i.e., a finite discrete population such as all the SAT scores administered in a given year). It can be shown that if $x \sim N(\mu, \sigma^2)$, then $z \sim N(0,1)$. In other words, if x has a normal distribution with average μ and variance σ^2, then z has a normal distribution with average 0 and variance 1.

Example: Given a normally distributed variable x with average 10 and standard deviation 2, calculated the z-score for $x = 9$. $z = \frac{x-\mu}{\sigma} = \frac{9-10}{2} = 0.5$

Note: A table of standard normal probabilities records values of z versus the area to the left of z under the the standard normal distribution function, $N(0,1)$. If $N(\mu, \sigma^2)$ were used, one would have to consider the infinite number of possible values for μ and σ^2.

Effect of changing the units of a variable on the mean

The mean of a variable X is specifically called the expected value of X. For a discrete variable X the expected value of X is the sum over all n data points of x_i times the probability P that the variable X will be equal to x_i . This is written as,

$$E[X] = \sum_{i=1}^{i=n} x_i\, P(X = x_i)$$

P for a discrete variable is also known as the relative frequency. If we change the units of X by a conversion factor a,

$$E[aX] = \sum_{i=1}^{i=n} ax_i\, P(aX = ax_i) = \sum_{i=1}^{i=n} ax_i\, P(X = x_i) = a \sum_{i=1}^{i=n} x_i\, P(X = x_i)$$

$E[aX] = aE[X]$.

Changing X by a conversion factor a multiplies the mean by a.

For a continuous variable X the expected value of X is the integral over the range of x of x times the probability P that the variable X will be equal to x. This is written as,

$$E[X] = \int_{x=x_{min}}^{x=x_{max}} xP(x)dx$$

P for a continuous variable is also known as the probability distribution function (pdf). If we change the units of X by a conversion factor a,

$$E[aX] = \int_{ax=ax_{min}}^{ax=ax_{max}} axP(ax)dax = \int_{x=x_{min}}^{x=x_{max}} axP(x)dx = a \int_{x=x_{min}}^{x=x_{max}} xP(x)dx$$

$E[aX] = aE[X]$. Changing X by a conversion factor a multiplies the mean by a.

Effect of changing the units of a variable on the variance

The variance of X is the expectation of the squared difference between X and the expectation of X. This is written as,$Var[X] = E([X - E(X)]^2) = E(X^2 - 2XE(X) + [E(X)]^2) = E(X^2) - 2E(X)E(X) + [E(X)]^2 Var[X] = E(X^2) - [E(X)]^2$

If we change the units of X by a conversion factor a,
$Var[aX] = E((aX)^2) - [E(aX)]^2 = a^2 E([X - E(X)]^2)$
$Var[aX] = a^2 Var[X]$. Changing X by a conversion factor a multiplies the variance by a^2.

Effect of changing the units on the mean, median, variance, range and IQR

After changing the units, the expectation of the mean $E[X]$ becomes $E[aX] = aE[X]$.

Changing X by a conversion factor a multiplies the mean by the conversion factor a.

The variance $Var[X]$ becomes $Var[aX] = a^2Var[X]$.

Changing X by a conversion factor a multiplies the variance by a^2.

Changing X by a conversion factor a multiplies the median by the conversion factor a.

Changing X by a conversion factor a multiplies the mode by the conversion factor a.

The interquartile range is calculated: $IQR = Q3 - Q1$. After changing the units, $aQ3 - aQ1 = a(Q3 - Q1) = a(IQR)$.

So, changing X by a conversion factor a multiplies the IQR by the conversion factor a. The $range = x_{max} - x_{min}$. After changing the units, $ax_{max} - ax_{min} = a(x_{max} - x_{min}) = a(range)$.

So, changing X by a conversion factor a multiplies the range by the conversion factor a.

Effect of changing the location of a variable on the mean

The mean of a variable X is specifically called the expected value of X. For a discrete variable X, the expected value of X is the sum over all n data points of x_i times the probability P that the variable X will be equal to x_i. This is written as,

$$E[X] = \sum_{i=1}^{i=n} x_i \, P(X = x_i)$$

P for a discrete variable is also known as the relative frequency. If we change the location of X by adding a,

$$E[X + a] = \sum_{i=1}^{i=n} (x_i + a) \, P(X + a = x_i + a) = \sum_{i=1}^{i=n} x_i \, P(X = x_i) + \sum_{i=1}^{i=n} a \, P(X = x_i)$$

$$E[X + a] = E[X] + a.$$

Changing the position of X by adding a, adds a to the mean.

For a continuous variable X the expected value of X is the integral over the range of x of x times the probability P that the variable X will be equal to x. This is written as,

$$E[X] = \int_{x=x_{min}}^{x=x_{max}} xP(x)dx$$

P for a continuous variable is also known as the probability distribution function (pdf). If we change the location of X by adding a,

$$E[X+a] = \int_{x-a=x_{min}+a}^{x-a=x_{max}+a} xP(x)dx + \int_{x-a=x_{min}+a}^{x-a=x_{max}+a} aP(x)dx$$

$E[X+a] = E[X] + a$. Changing the position of X by adding a, adds a to the mean.

Effect of changing the location of a variable on the variance

The variance of X is the expectation of the squared difference between X and the expectation of X. This is written as,$Var[X] = E([X - E(X)]^2) = E(X^2 - 2XE(X) + [E(X)]^2) = E(X^2) - 2E(X)E(X) + [E(X)]^2$
$Var[X] = E(X^2) - [E(X)]^2$

If we change the location of X by a in the positive direction,
$Var[X+a] = E((X+a)^2) - [E(X+a)]^2 = E(X^2 + 2aX + a^2) - ((E(X)^2 + 2aE(X) + a^2)$
$= E(X^2) + 2aE(X) + E(a^2) - E(X)^2 - 2aE(X) - a^2)$
$= E(X^2) + a^2 - E(X)^2 - a^2) = E(X^2) - E(X)^2)$
$Var[X+a] = Var[X]$.

Changing the position of X by adding a, does not change the variance.

Effect of changing the position of X by adding a on the mean, median, variance, range and IQR

After changing the position of X, the expectation of the mean $E[X]$ becomes $E[X+a] = E[X] + a$.

Changing the position of X by adding a, adds a to the mean.

The variance $Var[X]$ becomes $Var[X+a] = Var[X]$.

Changing the position of X by adding a, does not change the variance.

Changing the position of X by adding a, adds a to the median.

Changing the position of X by adding a, adds a to the mode.

The interquartile range is calculated: $IQR = Q3 - Q1$.

After changing the position of X by adding , $(Q3 + a) - (Q1 + a) = (Q3 - Q1) = IQR$.

So, changing the position of X by adding a, does not change the IQR.

The $range = x_{max} - x_{min}$.

After changing the position of X by adding a,
$(x_{max} + a) - (x_{min} + a) = (x_{max} - x_{min}) = range$.

So, changing the position of X by adding a, does not change the range.

Variables

There are two types of variables – categorical data and quantitative data. Sometimes categorical data is referred to as qualitative data. Quantitative data may be discrete or continuous.

Multiple distributions of categorical data may be graphically displayed using bar charts or pie charts.

Multiple distributions of quantitative data may be graphically displayed using histograms, dot plots, stem plots (also known as stem-and-leaf plots), box plots and cumulative frequency plots.

When using histograms, if the width of each bar is too wide, patterns in the shape can be lost. If the width is too narrow, misleading patterns may appear. The narrower the bar width, the more data is required to support conclusions about patterns.

Dot plots avoid this problem, and they will identify potential outliers, but they can only handle relatively small amounts of data.

Stem plots can have the same problem as histograms, plus they can only handle relatively small amounts of data.

A box plot will handle larger amounts of data, and it will identify the first, second and third quartiles, as well as outliers.

Stacked dot plots

Dot plots are useful for calling attention to clusters, gaps and outliers. It is useful for relatively smaller data sets, otherwise, histograms may be used. Similar to the effect the width of the bars in a histogram can have on the final appearance, the width of the dots can affect the appearance of a dot plot. Preferably, the width of the dots can be optimized by minimizing the difference between the observed and predicted data. One can visually estimate the optimum dot width.

When comparing dot plots from different sets of data, they are placed one above the other. It is preferable to optimize the dot width for both, since this can affect the appearance of the plots and this will be especially important when making comparisons.

Back-to-back stem plots

Stem plots are useful for calling attention to clusters, gaps and outliers. It is useful for relatively smaller data sets, otherwise, histograms may be used. Similar to the effect the width of the bars in a histogram and the width of dots in a dot plot can have on the final appearance, the number of digits of each datum assigned to the stems can affect the appearance of a stem plot. Preferably, the assignments can be optimized by minimizing the difference between the observed and predicted data. One can visually estimate the optimum dot width.

When comparing stem plots from different sets of data, they are placed side-by-side using a common stem. Generally speaking, two data sets with similar numbers of data points will be suitable for comparison.

Parallel box plots

Two or more sets of data may be compared using the descriptive technique of box plots by a parallel arrangement using a common axis on which to measure the variable. Each data set is labeled. Different box plots may be compared for the location of the median, the size of the interquartile range, high and low points, and the location of the median with respect to the center of the box which is a measure of skewness.

For example, the box plot of the daily value of a stock may be arranged in parallel with any number of stocks of similar industries to compare the median price (Q2), the high- and low-value and the volatility (as measured by the interquartile range). In another example, a set of parallel box plots can be used to examine the change of environmental carbon dioxide during the changing seasons.

Additional data that can be easily placed on a box plot are upper and lower bars on the whiskers indicating the 5th and 9th percentiles.

Double bar chart

A double (triple, etc.) bar chart is used to compare two (or more) sets of variables. The variables must be categorical data. The bar chart is preferred over the pie chart, because it is easier to view differences in the height of the bars than differences in the area of pie slices. The bar chart may be arranged so that the bars are all horizontal or all vertical. The different categorical variables are identified by using different fills for the bars. The fill is either a color or a black & white pattern, such as black, white, gray, stippled, stripes, etc.

A bar chart may also be used with bivariate data in which a categorical variable is paired with a quantitative variable. An example is a bar chart comparing the average income for different ethnic groups.

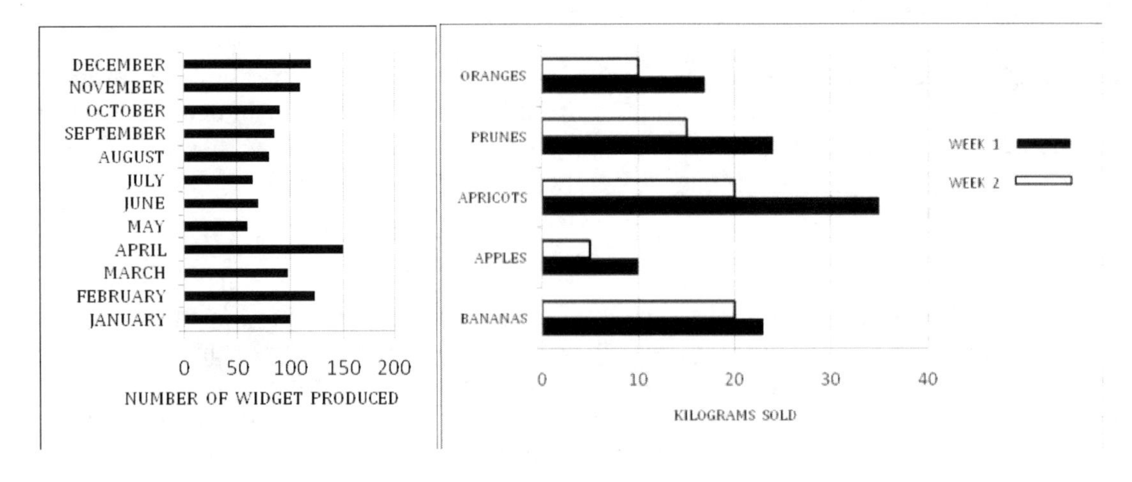

Cumulative frequency plots and Q-Q plots

From a cumulative frequency plot, the first quartile (Q1) is read off the x-axis for y=25%, the median (second quartile, Q2) is read off the x-axis for y=50%, the third quartile (Q3) is read off the x-axis for y=75%, and the interquartile range (IQR) which contains the middle 50% of all data points is Q3−Q1. When cumulative frequency plots from different data sets or distributions are placed side-by-side, all of these numerical measures may be compared. In addition, the Q1, Q2, Q3 and Q4 values of each distribution or sample set can be plotted against each other to form what is called a Q-Q plot. If such a Q-Q plot is a straight line, the two sample sets may have the same underlying distribution. One can even make a Q-Q plot of one of the sample sets against a theoretical distribution function like the standard normal distribution. If the resulting line is straight, it is likely that the sample set is normally distributed.

Recording bivariate data and the assignment of axes

Bivariate data is paired data where each datum is of the form (x, y). It is typically graphed as a scatter plot. The independent and dependent variables may be plotted on either axis. The scatter plot shows the relationship between the variables.

Central Limit Theorem

The second fundamental theorem of probability is the Central Limit Theorem (CLT). It states that as the sample size n of a variable X increases, the distribution of the sample average $\bar{x}$ approaches a normal distribution with mean μ (viz., $E(\bar{x}) = \mu$)and variance $\frac{\sigma^2}{n}$ (viz., $Var(\bar{x}) = \frac{\sigma^2}{n}$) irrespective of the distribution of the variable X.
Note: The X still comes from an unknown distribution, but $\bar{x}_n$ "converges in distribution" to a normal distribution. "Converging in distribution" does not mean $\bar{x}_n$ converges to a specific value; rather, it means the distribution of $\bar{x}_n$ converges to a normal distribution.

This can be re-stated, Z_n converges in distribution to $N(0,1)$, where,

$$Z_n = \frac{\bar{x} - \mu}{\sigma/\sqrt{n}}$$

and N is a normal distribution with average 0 and variance 1.
Proof: Recall, if a is a constant,
$E(ax) = aE(x), E(x \pm y) = E(X) \pm E(Y), E(a) = a, Var(ax) = a^2Var(x)$,
$Var(x \pm y) = Var(x) + Var(y), Var(a) = 0$.

$$E\left(\frac{\bar{x} - \mu}{\sigma/\sqrt{n}}\right) = \frac{1}{\sigma/\sqrt{n}}\{E(\bar{x}) - E(\mu)\} = \frac{1}{\sigma/\sqrt{n}}\{\mu - \mu\} = 0,$$

$$Var\left(\frac{\bar{x} - \mu}{\sigma/\sqrt{n}}\right) = \frac{1}{\sigma^2/n}\{Var(\bar{x}) + Var(\mu)\} = \frac{1}{\sigma^2/n}\{\sigma^2/n + 0\} = 1.$$

Sampling distribution of a sample proportion k/n with a sample space S={A,B}, k is the number of events A in the sample, and n is size of the sample

A sample proportion of a sample space S={A,B} is the fraction of the sample that is either event A. For example, if S={female, male}, a sample proportion could be the fraction of people that are female. The value k would distribute as a binomial distribution, $b(k, n, p)$

where k the number of females, n is the number of people in the sample, p is the probability that a person is female, and k/n is an estimate of p and is referred to as the sample proportion of females.

For a binomial distribution, $E(k) = np$ and $Var(k) = np(1-p)$, so recalling that $E(ax) = aE(x)$,
$E\left(\frac{k}{n}\right) = \frac{1}{n}E(k) = \frac{1}{n}np = p,$
and recalling that $Var(ax) = a^2Var(x)$,
$Var\left(\frac{k}{n}\right) = \frac{1}{n^2}Var(k) = \frac{1}{n^2}np(1-p) = \frac{p(1-p)}{n}.$

For two independent sample proportions,
$p_1 = \frac{k_1}{n_1}$ and $p_2 = \frac{k_2}{n_2}$,

$$E(p_1) = p_1, E(p_2) = p_2, Var(p_1) = \frac{p_1(1-p_1)}{n_1}, \ Var(p_2) = \frac{p_2(1-p_2)}{n_2}.$$

Z-scores and probabilities

The commonly used Z-scores and probabilities include the following:

Find the z for which the area to the left of it is 0.90 or 0.95 or 0.99.

Answers: 1.282, 1.645 and 2.326, respectively.

Find the z for which the area to the right of it is 0.90 or 0.95 or 0.99.

Answers: Same as above because of symmetry, 1.282, 1.645 and 2.326, respectively.

Find the z for which the area between $+z$ and $-z$ is 0.90 or 0.95 or 0.99.

Answers: 1.645, 1.960 and 2.576, respectively.

The area between $z = \pm 1$ is 0.6826. Find $\pm x$ $(x = z\sigma + \mu)$.

The area between $z = \pm 2$ is 0.9544. Find $\pm x$.

The area between $z = \pm 3$ is 0.9974. Find $\pm$.

Probability

Probability

Probability is expressed as a relative frequency. For a discrete distribution, the probability that a variable X has certain values is defined as the sum of relative frequencies for each of those certain values of X,

$$\sum_{i=1}^{i=N} \delta_i f(x_i)$$

where $f(x_i)$ has the property that

$$\sum_{i=1}^{i=N} f(x_i) = 1$$

and $f(x_i)$ is the relative frequency that x_i occurs in the population, and $\delta_i = 1$ if x_i is one of the values of X whose relative frequency you wish to determine, otherwise $\delta_i = 0$.

Example: If two dice are tossed, the variable X is the outcome (sum of the two dice),

DIE #1	DIE #2	POSSIBLE OUTCOMES FOR X
1	1 2 3 4 5 6	2 3 4 5 6 7
2	1 2 3 4 5 6	3 4 5 6 7 8
3	1 2 3 4 5 6	4 5 6 7 8 9
4	1 2 3 4 5 6	5 6 7 8 9 10
5	1 2 3 4 5 6	6 7 8 9 10 11
6	1 2 3 4 5 6	7 8 9 10 11 12

Calculate the relative frequency of each outcome,

X	2	3	4	5	6	7	8	9	10	11	12
RELATIVE FREQUENCY	1/36	2/36	3/36	4/36	5/36	6/36	5/36	4/36	3/36	2/36	1/36

Calculate the probability, or sum of the relative frequencies, that X<8,

$$\sum_{i=1}^{i=11} \delta_i f(x_i) = \frac{1}{36} + \frac{2}{36} + \frac{3}{36} + \frac{4}{36} + \frac{5}{36} + \frac{6}{36} = \frac{21}{36} = \frac{7}{12}$$

For a continuous distribution, the probability is the integral of the probability distribution function (pdf) over the range of values for the variable whose probability one wishes to determine.

$$\int_A^B p(x)\,dx + \int_C^D p(x)\,dx + \cdots$$

Due to the definition of the integral, the relative frequency of a specific value for x is zero because, $\lim_{\Delta x \to 0} p(x)\Delta x = 0$. However, the relative frequency or probability is generally non-zero when integrated over a range of possible values for x.

The pdf has the property that when integrated over all possible values of x,
$\int p(x)\, dx = 1$
Example: The probability that variable X is greater than A and less than B is 1/2. This is written,

$$\Pr(A < X < B) = \int_A^B p(x)dx = 1/2.$$

Example: The probability that variable X is less than A and greater than B is 1/2. The domain for $p(x)$ is $[0,\infty)$. This is written,

$$\Pr(X < A, X > B) = \int_0^A p(x)dx + \int_B^\infty p(x)dx = 1/2.$$

IID variable and the Law of Large Numbers

An independent and identically distributed (iid) variable means one that is selected from a population independently from all the other selections, and that each selection of the variable comes from the same distribution.

The first fundamental theorem of probability is the Law of Large Numbers. The Law of Large Numbers states that the result of an experiment approaches the expected value (or sample mean) as the number of trials increases. For example, after tossing a coin 100 times, the outcome, or probability, approaches ½.

The consequence of the "Law of Large Numbers" is that it justifies the assumption that the expected value of a random variable(viz., the average) will continue to be approached more closely as more samples are included in the estimate.

Relative fraction

Chebyshev's inequality may be re-formulated to state that the relative fraction of the possible values of a variable with an unknown distribution that lie within $\pm k$ standard deviations of the mean is at least $(1-\frac{1}{k^2})$. The Chebyshev's inequality is written,
$Pr(|X-\mu| \geq k\sigma) \leq \frac{1}{k^2}$,

So, to re-formulate, we observe that $Pr(|X-\mu| \leq k\sigma) = 1 - Pr \geq (1-\frac{1}{k^2})$.
Given $|X-\mu| \leq k\sigma$,
if $|X-\mu| = X-\mu$ then $X-\mu \leq k\sigma$ and $X \leq \mu + k\sigma$.
if $|X-\mu| = -(X-\mu)$ then - $(X-\mu) \leq k\sigma$ and $\mu - k\sigma \leq X$,
and so for both cases, $\mu - k\sigma \leq X \leq \mu + k\sigma$.

So Chebyshev's inequality may be re-formulated,
$Pr(\mu - k\sigma \leq \leq \mu + k\sigma) \geq (1-\frac{1}{k^2})$

Example: Suppose that a manufacturer has assembled 10,000 units of a metal rod whose length has standard deviation of 100. With no knowledge of what the distribution looks like, we can let $k = 2$ and state that at least $(1 - \frac{1}{2^2})$ or $\frac{3}{4}$ of the rods are within ± 2 standard deviations of the mean. Using the above notation, $Pr(\mu - 2\sigma \leq X \leq \mu + 2\sigma) \geq \frac{3}{4}$.

Set

A set is a collection of elements.

Example: X={a, b, c, d}

In probability theory, the elements are known as events or outcomes

The complement A^C of a set A relative to set B is the set of elements in B but not in A.

Example: A={a, b, c, d}, B={a, b, c, d, e, f}, A_C = {e, f}.

The union of A and B is the set containing all the distinct elements of A or B.

Example: A={a, b, c, d}, B={a, b, c, d, e, f}, A∪B={ a, b, c, d, e, f}.

The intersection of A and B is the set containing all the elements that are common to A and B.

Example: A={a, b, c, d}, B={a, b, c, d, e, f}, A∩B={ a, b, c, d}.

Probability space, sample space, set of events or outcomes

A probability space, (S, E, P), consists of a sample space S, a set of events E, and a probability measure P.

A sample space, S, is the set of all possible outcomes or events.

Example: In a single coin toss, S={head, tail}.

Example: In the toss of one die, S={1,2,3,4,5,6}.

The set of events or outcomes E is a subset of the sample space, including the sample space itself, such that the complements and all possible countable unions of the subsets are also subsets of E. Note that a set A may be referred to as an event, but an event is generally a subset of the sample space. The subset may contain one or more elements. For example, the event A may be the arrival of two ships (a, b) in a fleet of three (a, b, c). In this case, the event A={{a, b}, {b, c}, {a, c}}.

Probability measure in a probability space

Suppose you will roll a die. The sample space is the list of all possible outcomes, i.e., {1, 2, 3, 4, 5, 6}. Also, suppose that you wish to know the probability that you will roll an even number. There are three possible even numbers. The elements in the event are namely {2, 4, 6}. Therefore, the probability of rolling an even number is 3/6 or ½.

The probability measure P for a discrete distribution assigns a probability for each discrete event.

Construct an event set E from the sample space S for each of the following random samples: (i) a single coin toss, (ii) two rolled dice, and (iii) two coin tosses

The sample space S is determined by the situation. E must contain S and its relative complement ∅. The probability that there will be no outcome is zero, P(∅)=0. The probability that the outcome will be some element of the sample space is 1, P(S)=1.

Example (i): A coin with a head (h) and a tail (t) is tossed once.

S={h, t},

E{h} Probability of head is 1/2.

Example (ii): sum of two dice rolled.

S={2, 3, 4, 5, 6, 7, 3, 4, 5, 6, 7, 8, 4, 5, 6, 7, 8, 9, 5, 6, 7, 8, 9, 10, 6, 7, 8, 9, 10, 11, 7, 8, 9, 10, 11, 12}, Note. There are 36 elements in the sample space, because the total number of possible sums is found by multiplying the possibilities on each die, i.e., 6 × 6.

E={a sum less than 6}

The probability of a sum less than 6 is 10/36 or 5/18.

Example (iii): A coin is tossed twice. S = {(head, head), (head, tail), (tail, tail), (tail, head)} Notice there are 4 elements in the sample space, since there are two possibilities for each coin, and 2 × 2 = 4.

E = {two heads}; the probability of two heads is 1/4.

Create an example of a Cartesian product sample space and calculate the probability of an event that is a subset of the sample space

Example: A company has bid two contracts. The sample space, possible outcomes, and probabilities are as follows,

$S_1 \times S_2$={win×win,lose×lose,win×lose,lose×win},

where the subscript on the S refers to contract #1 and contract #2, and win×lose means the company wins contract #1 and loses contract #2.

E= {win, win}; the probability of the aforementioned event is ¼. The company has a ¼ chance that it will win both contracts.

Union and intersection of two sets of outcomes

Let A and B each be a set of elements or outcomes. Since the application is probability theory, we will use the term "outcomes".

The union (symbol ∪) of two sets is the set of elements found in Set A or Set B.

Example: A={2,3,4}, B={3,4,5}, A∪B={2,3,4,5}. Note that the outcomes 3 and 4 appear only once in the union.

The intersection (symbol ∩) of two sets is the set of outcomes common to both sets.

Example: A={2,3,4}, B={3,4,5}, A∩B={3,4}.

Union and intersection commute: $A \cup B = B \cup A; A \cap B = B \cap A$

The union of sets is left and right distributive over intersection,

left distributive: $A \cup (B \cap C) = (A \cup B) \cap (A \cup C)$,

right distributive: $(B \cap C) \cup A = (B \cup A) \cap (C \cup A)$

The intersection of sets is left and right distributive over union.

left distributive: $A \cap (B \cup C) = (A \cap B) \cup (A \cap C)$,

right distributive: $(B \cup C) \cap A = (B \cap A) \cup (C \cap A)$

Probability of the complement of a set of events

For a probability space (S, E, P) with sample space S, set of events E, and probability measure P that assigns a probability to each event, let A be one or more events contained in the set of events E. Then the complement of A relative to E is written A^C, and $P(A^C)=1-P(A)$.

Example: Given the following probability space (S,E,P),

S={win contract, lose contract},
E={win contract},
P(win contract)=1/2,
calculate $P([\text{win contract}]^C)$.
$P([\text{win contract}]^C) = 1 - P(\text{win contract}) = 1/2$.

Example: A company has bid two contracts. Given the following probability space (S,E,P),

$S_1 \times S_2$={win×win,lose×lose,win×lose,lose×win}, where the subscript on the S refers to contract #1 and contract #2, and win×lose means the company wins contract #1 and loses contract #2,E{lose×win}; the probability of (lose, win) is ½.

calculate $P([\text{lose}\times\text{win}]^C)$, the probability that the company will not win at least one contract and lose at least one contract.

$P([\text{lose}\times\text{win}]^C) = 1 - 0..500 = 0.50$.

General addition rule

$P(A \cup B) = P(A) + P(B) - P(A \cap B)$,where P(A) is the probability the event(s) in A will occur, P(B) is the probability the event(s) in B will occur, $P(A \cup B)$ is the probability the event(s) in the union of A and B will occur (probability that A or B will occur), and $P(A\cap B)$ is the probability the event(s) in the intersection of A and B will occur (probability that A and B will occur).

Suppose you wish to determine the probability that you roll a 3 when rolling a single die or get a head when tossing a coin. Since you see the word "or", you know you are looking at the probability that the events in set A or set B will occur. Therefore, you will use the formula shown above, i.e., $P(A \cup B) = P(A) + P(B) - P(A \cap B)$. Given that the roll of a die constitutes events in Set A and the tossing of a coin constitutes events in Set B, you can substitute probabilities into the addition formula and get: $\frac{1}{6} + \frac{1}{2} = \frac{2}{3}$. In this case, the events are mutually exclusive, and thus there are not any events in the intersection.

Suppose you wish to determine the probability that you roll a 3 or you roll an odd number when rolling a single die. These events are not mutually exclusive because 3 is actually an odd number. In this case, you get: $\frac{1}{6} + \frac{3}{6} - \frac{1}{6} = \frac{1}{2}$. Note. The 1/6 represents the intersection of the events in Sets A and B.

Probability of intersecting event(s) A∩B and the probability of conditional event(s) B|A

$P(B|A) = \frac{P(B \cap A)}{P(A)}$where P(A) is the probability the event(s) in A will occur, $P(B \cap A)$ is the probability the event(s) common to both B and A will occur, and P(B|A) is the probability the event(s) in B will occur on the condition that the event(s) in A occurred. This is known as conditional probability.

Example: Suppose you are asked to draw a colored marble out of a bag. There are 4 green marbles, 10 red marbles, 8 yellow marbles, and 7 blue marbles. You would like to know the probability of first drawing a green marble and then drawing another green marble, with the condition that you will not replace the marble. Thus, you would write the conditional probability as P(drawing a second green marble | first drawing a green marble). On the first draw, you have a probability of 4/29 of drawing a green marble. Since you will not replace the marble, the new event size is 3 and the sample size stays the same. Thus, the probability of drawing a second green marble after drawing a green marble is 3/29.

Difference between $P(A \cap B)$ and $P(A|B)$ in terms of the sets A, B and A∩B, assuming the sample space is {A,B}

Assuming the sample space is {A,B}, then only those two events can occur, $P(A) + P(B) = 1$, and the following definitions are valid.

The definition of P(A∩B)is,

$$P(A \cap B) = \frac{A \cap B}{A + B}.$$

In words,

$$(probability\ both\ A\ and\ B\ occur) = \frac{\#\ of\ times\ both\ A\ and\ B\ occur}{\#\ of\ times\ A\ occurs + \#\ of\ times\ B\ occurs}.$$

The definition of P(A|B) is,

$$P(A|B) = \frac{A \cap B}{B}.$$

In words,

$$(probability\ A\ occurs\ on\ the\ condition\ that\ B\ has\ also\ occured)$$
$$= \frac{\#\ of\ times\ both\ A\ and\ B\ occur}{\#\ of\ times\ B\ occurs}.$$

Note: $\frac{A \cap B}{A + B} \neq \frac{P(A \cap B)}{P(A) + P(B)}$,

because $P(A) = \frac{\#\ of\ times\ A\ occurs}{\#\ of\ times\ A\ occurs + \#\ of\ times\ B\ occurs}$.

Derive Bayes Theorem

From the definition of $P(A|B)$ and $P(B|A)$,

$$P(A|B) = \frac{P(A \cap B)}{P(B)}, P(B|A) = \frac{P(A \cap B)}{P(A)}.$$

Rearranging these two equations and combining them produces,
$P(B)P(A|B) = P(A \cap B) = P(B|A)P(A)$.
Omitting $P(A \cap B)$ gives,
$P(B)P(A|B) = P(B|A)P(A)$.

Rearranging gives Bayes Theorem,

$$P(A|B) = \frac{P(B|A)P(A)}{P(B)}.$$

Example: A drug test given employees is 98% accurate (the test is positive for 98% of drug users). The test gives a false positive only 1% of the time (the test is positive for 1% of people who are <u>not</u> drug users).

D = employee is a drug user, E = employee is <u>not</u> a drug user, + = test is positive,
P(+|D) = 0.98, P(+|E) = 0.01, Historically P(D)=0.007, so P(E)=0.993.

The prob. of a positive test is NOT the prob. of a <u>true</u> positive test and prob. of a <u>false</u> positive test; rather, it is the probability of the following two intersections,
$P(+) = P(+ \cap D, + \cap E) = P(+ \cap D) + P(+ \cap E)$, and by definition of conditional probability,$P(+) = P(+|D)P(\quad) + P(+|E)P(E) = 0.98(0.007) + 0.01(0.993) = 0.01679$.

From Bayes Theorem, the prob. that an employee who tests positive actually uses drugs is,

$$P(E|+) = \frac{P(+|E)P(E)}{P(+)} = \frac{0.01(0.993)}{0.01679} = 0.59.$$

Multiplication rule

The probability of the intersection of A and B is the probability of A times the probability of B. This is written, $P(A \cap B)=P(A)P(B)$.

This is known as the multiplication rule.

In using the multiplication rule, suppose you are examining two independent events. In this case, you simply multiply two probabilities to get the probability of the intersection of the probabilities. Likewise, if you start with a given intersection and the value is the same as the product of the probability of set A and set B, then the events are independent. Let's look at an example. Suppose you want to determine if rolling a 4 on a single die and getting a tail when tossing a coin are two independent events. Well, the probability of rolling a 4 is 1/6, while the probability of getting tails is ½. The product of these two events is 1/12; the probability of the intersection is also 1/12 because there are 12 possibilities in the sample space and only one way of getting a 4 and tossing a tail. Therefore, the two events are independent.

Multistage probability calculations

When event A is not conditional on B nor vice-versa, $P(A \cap B)=P(A)P(B)$; otherwise, $P(A \cap B)=P(A|B)P(B)=P(B|A)P(A)$.

Example: Every Saturday Mr. Jones goes to either Bank #1 or Bank #2 to deposit checks. Event B1 is Mr. Jones going to Bank #1 and event B2 is Mr. Jones going to Bank #2. Event $C_{>50}$ is Mr. Jones having checks totaling $50 or more. On average, $P(C_{>50})=0.8$.

Mr. Jones walks to either bank with equal probability.

$C_{>50}$ is clearly independent of B1 or B2, because Mr. Jones has the checks before he decides which bank to go to. The reasoning is logical not mathematical.

Recall, B1={going to bank#1 with C>50, going to bank#1 with C<50} and
$C_{>50}$ ={going to bank#1 with C>50, going to bank#2 with C>50} and
$B1\cap C_{>50}$={going to bank#1 with C>50}.

So the probability of event B1 and event $C_{>50}$ is, $P(B1\cap C_{>50}) = P(B1)P(C_{>50}) = 0.5(0.8) = 0.4$.

When event A is not conditional on B nor vice-versa, $P(A \cap B)=P(A)P(B)$; otherwise, $P(A \cap B)=P(A|B)P(B)=P(B|A)P(A)$.

Example: Every Saturday Mr. Jones goes to either Bank#1 (B1) or Bank#2 (B2) to deposit checks. Event $C_{>50}$ is Mr. Jones having checks totaling $50 or more, and $P(C_{>50}) = 0.8$.

Event R is rain on Saturday, the complement of R is sunshine and $P(R)=0.2$.

On sunny days [event R^C] Mr. Jones walks to either bank with equal probability [$P(B1|R^C)= P(B2|R^C)=0.5$], but if it is raining, Mr. Jones goes to Bank#1 with probability 0.1 and to Bank#2 with probability 0.9 [$P(B1|R)=0.1$, $P(B2|R)=0.9$].

The probability of B1 is the probability of (B1 and rain) plus the probability of (B1 and sun), $P(B1)=P(B1\cap R)+P(B1\cap R^C)$.
$P(B1)=P(B1|R)P(R)+P(B1|R^C)P(R^C)$
$P(B1)=0.1(0.2)+0.5(1-0.2)=0.42$.

But $P(B1\cap R)=0.1\neq P(B1)P(R)=0.42(0.2)=0.084$, so B1 and R are not independent. The reasoning is mathematical.

$C_{>50}$ is clearly independent of B1 or B2, because Mr. Jones has the checks before he decides which bank to go to. The reasoning is logical.

The probability that Mr. Jones goes to bank#1 with checks totaling $50 or more is $P(B1\cap C_{>50})= P(B1)P(C_{>50})=0.42(0.8)=0.336$

Similarly, $P(B2\cap C_{>50})= P(B2)P(C_{>50})=0.58(0.8)=0.464$.

How many ways can you arrange *k* indistinguishable balls in *n* distinguishable bins?

Consider how many ways we can arrange k indistinguishable balls in n distinguishable bins, where each bin cannot hold more than one ball. In the figure below are 3 balls and 5 bins. The balls cannot be numbered because they are supposed to be indistinguishable, however, the bins can be numbered, because they are distinguishable.

There are 5 ways to add the 1st ball to one of the 5 bins. There are 4 ways to add the 2nd ball to the remaining 4 bins, and 3 ways to add the 3rd ball to the remaining 3 bins.

$(5)(4)(3) = \frac{(5)(4)(3)(2)(1)}{(2)(1)}$, and, in general, $\frac{n!}{(n-k)!}$. However, contained in the 60 configurations are identical configurations, because the balls are indistinguishable. The figure above, can be formed in 3! ways, so we must divide our general expression by $k!$. The number of distinguishable ways to arrange k balls in n bins is $\frac{n!}{k!(n-k)!}$.

Binomial distribution

The number of ways to arrange k indistinguishable balls in n distinguishable bins is $\frac{n!}{k!(n-k)!}$. The probability that a ball will be placed in a bin is p, and, therefore, the probability that a ball will not be placed in a bin is $(1-p)$. The probability of any configuration of 3 balls in 5 bins is $p^3(1-p)^2$ and in general, the probability of any configuration of k balls in n bins is $p^k(1-p)^{(n-k)}$. The probability of all 10 distinguishable configurations of 3 balls in 5 bins is $10p^3(1-p)^2$. In general, the probability of all distinguishable configurations of k balls in n bins is the number of all configurations times their probability,

$$\frac{n!}{k!\,(n-k)!}p^k(1-p)^{n-k}$$

This is the binomial distribution $\mathrm{b}(k,n,p)$ where n and p are the parameters, k is the variable in the probability distribution function $b(k,n,p) = \frac{n!}{k!(n-k)!}p^k(1-p)^{n-k}$ and $k \sim \mathrm{b}(k,n,p)$ ($\sim$ means "distributes as").

The process associated with a binomial distribution is also thought of as n independent trials and k successes, each trial with the same probability p of success. Recording the number of successes k each time n trials performed will reveal that $k \sim \mathrm{b}(k,n,p)$.

To check the integrity of circuit boards before they are loaded with expensive components, a manufacturer cuts a portion of the circuit board containing n=5 thru-holes. The sample is exposed to molten solder several times to induce damage to susceptible through-holes. The number of holes that fail the test is k. Historically, the probability of a hole failure is $p = 0.25$.

The process associated with a binomial distribution is often thought of as n independent trials and k successes, each with the same probability of occurrence. In this example, each of the n thru-holes is an independent trial with the same probability p of success or passing without failure, and k is the number of holes that pass without failure.The expected value of hole failure, $\mathrm{E}(k)$, and variance?

$$E(k) = \sum_{k=0}^{k=5} k\frac{n!}{k!\,(n-k)!}p^k(1-p)^{n-k}$$

$$= 0\frac{5!}{0!\,5!}0.25^0(0.75)^5 + 1\frac{5!}{1!\,4!}0.25^1(0.75)^4 + 2\frac{5!}{2!\,3!}0.25^2(0.75)^3 + 3\frac{5!}{3!\,2!}0.25^3(0.75)^2 + 4\frac{5!}{4!\,1!}0.25^4(0.75)^1 + 5\frac{5!}{5!\,0!}0.25^5(0.75)^0$$

$$= 0(0.75)^5 + 1(5)0.25^1(0.75)^4 + 2\frac{20}{2}0.25^2(0.75)^3 + 3\frac{20}{2}0.25^3(0.75)^2 + 4(5)0.25^4(0.75)^1 + (5)0.25^5$$

$$= 5(0.25)\text{=}1.25$$

In general, $E(k) = np$ and $Var \equiv E\left[\left(k - E(k)\right)^2\right] = np(1-p)$.

In this example, $E(k) = np = 5(0.25) = 1.25, Var = np(1-p) = 5(0.25)(0.75) = 0.9375$.

The average and standard deviation are 1.25 and 0.968. We cannot write 1.25$\pm$0.97 because the distribution is not symmetrical about the average.

Multinomial distribution

A multinomial distribution is a generalization of the binomial distribution. The distribution in a survey of a population may be binomial if there are only two possible outcomes, but it may be multinomial if there are more than two possible outcomes. For a multinomial distribution, the average and variance for the ith outcome in a total of τ outcomes is,

$E(\bar{k}_i) = np_i, Var(\bar{k}_i) = np_i(1-p_i)$, where $\sum_{i=1}^{i=\tau} p_i = 1$.
The estimate $\hat{p}_i$ for each p_i is $\hat{p}_i = {}^{k_i}/_{n_i}$.

The estimates for the means and variances are,

$$\frac{k_i}{n_i} = \hat{p}_i,\ SEM(\hat{p}_i) = \sqrt{\frac{\hat{p}_i(1-\hat{p}_i)}{n_i}}.$$

Example: A survey of 1000 people for the top three choices for senatorial candidate in the Democratic primary gave the following results, for candidates A, B and C, $k_A = 300, k_B = 600$, and $k_C = 100$. $\hat{p}_A = 0.300, \hat{p}_B = 0.600, \hat{p}_C = 0.100$.

$$SEM_A = \sqrt{\frac{0.3(1-0.3)}{300}} = 0.026, SEM_B = \sqrt{\frac{0.6(1-0.6)}{600}} = 0.020, \text{ and } SEM_C = \sqrt{\frac{0.1(1-0.1)}{100}} = 0.0300.$$

Candidate A: 30.0%±2.6%, Candidate B: 60.0%±2.0%, Candidate C: 10.0%±3.0%.

Negative binomial distribution in terms of arranging r indistinguishable balls in $(\mathbf{w+r})$ distinguishable bins connected in sequence wherein the probability of placing a ball in any bin is p (and, consequently, the probability of not placing a ball in any bin is $(\mathbf{1-p})$.)

First create $(w+r)$ distinguishable bins connected in sequence and place a ball in the last bin. If the number of distinguishable ways to arrange k balls in n bins is

$$\frac{n!}{k!\,(n-k)!},$$

then the number of distinguishable ways to arrange the remaining $(r-1)$ balls in the remaining $(w+r-1)$ bins is

$$\frac{(w+r-1!}{(r-1)!\,(w)!}.$$

If each of the above configurations has r bins with balls and w bins with no balls, the probability of each configuration is $p^r(1-p)^w$. The negative binomial distribution function will then be the number of configurations times their probability,

$$b_{neg}(w,r,p) = \frac{(w+r-1)!}{(r-1)!\,w!}p^r(1-p)^w.$$

The negative binomial distribution function has two fixed parameters r and p, and one variable w.

$$E(w) = \frac{r(1-p)}{p},\ Var(w) = \frac{r(1-p)}{p^2}.$$

Example of negative binomial distribution

If a ballplayer has a 1/3 probability of hitting a fast ball, what is the probability of 2 misses before he get 3 hits? What is the expected value (average) and variance for the number of misses before he gets 3 hits?

In the context of balls and bins, where a ball in a bin represents the ballplayer hitting the ball, and an empty bin represents a miss, the following are the six possibilities,

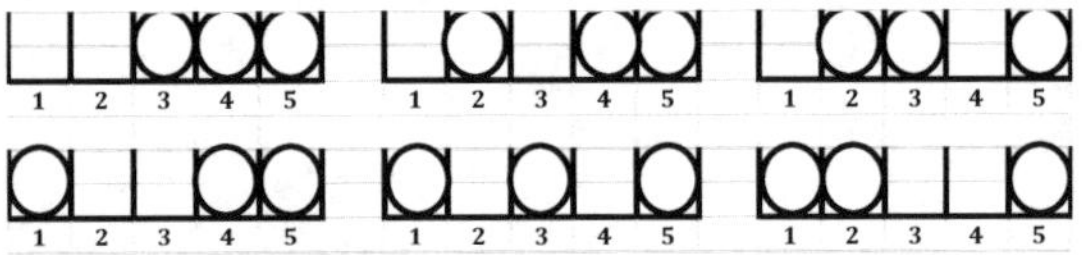

1 2 3 4 5 1 2 3 4 5 1 2 3 4 5

1 2 3 4 5 1 2 3 4 5 1 2 3 4 5

$w = 2, r = 3, p = 1/3$

$$b_{neg}(w,r,p) = \frac{(w+r-1)!}{(r-1)!\,w!}p^r(1-p)^w = \frac{(2+3-1)!}{(3-1)!\,2!}(0.35)^3(1-0.35)^2 = 6(0.018)$$
$$= 0.109.$$

$$E(w) = \frac{r(1-p)}{p} = \frac{3(1-1/3)}{1/3} = 6,\ Var(w) = \frac{r(1-p)}{p^2} = \frac{3(1-1/3)}{(1/3)^2} = 18.$$

The average and standard deviation are 6 and 4.2 . We cannot write 6 ± 4.2 because the distribution is not symmetrical about the average.

Geometric distribution in terms of $(z-1)$ bins in sequence with one ball in the zth bin

Using the process of placing no balls in any of the first $(z-1)$ bins connected in sequence and one ball in the zth bin where the probability of placing a ball in any bin is p (and, consequently, the probability of not placing a ball in any bin is $(1-p)$), it is clear there is only one possible configuration. For example, the configuration for z=5,

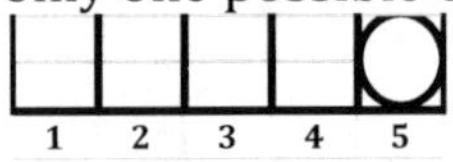

1 2 3 4 5

The probability of the above configuration is $(1-p)^4p$. The geometric distribution function will then be the number of possible configurations (one) times its probability,

$g\,(z,p) = (1-p)^{z-1}p.$

The geometric distribution function has one fixed parameter p and one variable z.

$$E(z) = \frac{1}{p},\ Var(z) = \frac{(1-p)}{p^2}.$$

Example of geometric distribution

If a ballplayer has a 0.35 probability of hitting a fast ball, what is the probability of 2 misses before he gets a hit? What is the expected value (average) and variance for the number of swings until he gets a hit?

$z = 3, p = 0.35$

$g\,(z,p) = (1-p)^{z-1}p = (1-0.35)^{3-1}(0.35) = 0.148.$

$$E(z) = \frac{1}{p} = \frac{1}{0.35} = 2.86,\ Var(z) = \frac{(1-p)}{p^2} = \frac{(1-0.35)}{0.35^2} = 5.3.$$

The average and standard deviation are 2.9 and 2.3. We cannot write 3 ± 2.3 because the distribution is not symmetrical about the average.

Note: Although one cannot have a fraction of a hit or miss, as the number of trials increases, the expected value and variance come closer to an integer value. For example, in a Bernoulli trial (a single event), if the probability of a coin toss for heads is 0.35, it might not be until there are 100 tosses before you will see closer to 35% heads.

Hypergeometric distribution

The hypergeometric distribution, $H(k, B, m)$, can be thought of as the distribution of k, the number of white balls selected out of a sample of n balls from an urn containing a total of N balls of which m are black and $(N - m)$ white.

If $H(k, B, m)$ is the distribution of k it is also the probability of selecting k black balls.

[Note: Summing up the probability over all possible values of k will equal unity.]

$$H(k, B, m) = \frac{(\#\ of\ ways\ to\ select\ k\ black\ balls)x(\#\ of\ ways\ to\ select\ (n-k)\ white\ balls)}{(\#\ of\ possible\ samples\ of\ size\ n\ drawn\ without\ replacement\ from\ N\ balls)},$$

$$(\#\ of\ possible\ samples\ of\ size\ n\ drawn\ without\ replacement\ from\ N\ balls) = \frac{N!}{(N-n)!\,n!} = \binom{N}{n},$$

$$(\#\ of\ ways\ to\ select\ k\ black\ balls) = \frac{m!}{(m-k)!\,k!} = \binom{m}{k},$$

$$(\#\ of\ ways\ to\ select\ (n-k)\ white\ balls) = \frac{(N-m)!}{(N-m-n+k)!\,(n-k)!} = \binom{N-m}{n-k}.$$

$$H(k, B, m) = \frac{\binom{m}{k}\binom{N-m}{n-k}}{\binom{N}{n}}$$

$\binom{N}{n}$ term in the hypergeometric distribution

$$H(k, B, m) = \frac{\binom{m}{k}\binom{N-m}{n-k}}{\binom{N}{n}}.$$

Inside the urn one can mathematically construct N distinguishable bins for the purpose of emphasizing that in order to count the number of possible selections one must "see" where each ball comes from. Each bin contains one ball. Changing the order of selection of a sample does not create a new sample.

$$(\#\ of\ possible\ samples\ of\ size\ n\ drawn\ without\ replacement\ from\ N\ balls) = \frac{N!}{(N-n)!\,n!} = \binom{N}{n},$$

Recall that $N!/(N-k)!$ is the number of ways to select n balls from N distiguishable bins but one must divide by $n!$ because one does not distinguish the order of selection.

Example: Demonstrate

$$(\#\ of\ possible\ samples\ of\ size\ n\ drawn\ without\ replacement\ from\ N\ bals) = \frac{N!}{(N-n)!\,n!} = \binom{N}{n}.$$

1 2 3 4 Let $N = 4, n = 2$.

12 ways to select two balls, from 4 distinguishable bins, where 13 means the balls from bins 1 and 3: 12, 13, 14 , 21, 23, 24, 31, 32, 34, 41, 42, 43

Half the selection are only different by the order of selection, so we divide 12 by 2 to get 6 possible samples of size 2 drawn without replacement from 4 and irrespective of sequence.

$$\frac{N!}{(N-n)!\,n!} = \frac{4!}{(4-2)!\,2!} = \frac{4*3*2}{(2)2} = 6,$$

and the formula is confirmed.

Binomial vs. geometric distributions

The process associated with a binomial distribution may be thought of as n independent trials each with probability p of a success, k is the number of successes, and

$$k \sim b(k,n,p) = \frac{n!}{k!\,(n-k)!} p^k (1-p)^{n-k}.$$

The binomial distribution function has two fixed parameters n and p, and one variable k.
$E(k) = np,\ Var(k) = np(1-p).$

The process associated with a geometric distribution may be thought of as independent trials each with probability p of a success, z is the number of trials until the first success, and

$z \sim g(z,p) = (1-p)^{z-1} p.$

The geometric distribution function has one fixed parameter p and one variable z.

$$E(z) = \frac{1}{p},\ Var(z) = \frac{(1-p)}{p^2}.$$

Geometric vs. negative binomial distributions

The process associated with a geometric distribution may be thought of as independent trials each with probability p of a success, z is the number of trials until the first success, and

$z \sim g(z,p) = (1-p)^{z-1} p.$

The geometric distribution function has one fixed parameter p and one variable z.

$$E(z) = \frac{1}{p},\ Var(z) = \frac{(1-p)}{p^2}.$$

The process associated with a negative binomial distribution may be thought of as independent trials each with probability p of a success, w is the number of failures before r successes, and

$$w \sim b_{neg}(w,r,) = \frac{(w+r-1)!}{(r-1)!\,w!} p^r (1-p)^w.$$

The negative binomial distribution function has two fixed parameters r and p, and one variable w.

$$E(z) = \frac{r(1-p)}{p},\ Var(z) = \frac{r(1-p)}{p^2}.$$

Binomially distributed variable simulated using a Bernoulli random number generator

Below are zeros and ones generated with a Bernoulli random number generator with p=1/2.

1	1	1	1	1	0	1	0	0	1
1	1	0	1	1	0	1	0	0	0
1	1	0	1	1	0	0	1	0	0
0	0	1	0	0	1	0	1	1	1
1	1	0	1	0	0	1	0	0	0
0	0	1	0	0	1	0	1	1	1
1	1	0	1	1	1	1	0	0	0
1	0	1	0	0	0	0	1	1	1
0	1	0	1	1	1	1	0	1	0
1	0	0	1	1	0	0	1	0	0

If the probability of a thru-hole failing is 50% (p=1/2) on a circuit board after five cycles of solder shock testing, can a reasonable estimate of this failure rate be made by examining 5 holes? 10 holes? 30 holes?

From the Bernoulli generated table on the left, counting the percentage of 1's from left to right on the top row, for the first 5 numbers the percentage of 1's is 100%, the first 10 is 70%, the first 30 is 57%. Clearly, 30 holes should be examined.

Geometrically distributed variable simulated using a Discrete Uniform random number generator

Below is a table of numbers 0,1,2,3,4,5,6,7,8,9 generated with a Discrete Uniform random number generator.

5	4	2	6	6	0	9	1	3	2	5	0
9	7	5	6	3	1	3	0	6	0	0	1
1	1	8	4	9	7	2	3	2	6	0	4
3	4	7	6	0	1	0	5	9	6	0	9
6	1	9	0	1	1	6	9	1	6	1	6
7	9	2	7	6	1	6	2	1	1	8	1
0	2	2	2	1	3	1	7	2	6	9	6
6	3	8	7	1	2	7	7	7	8	3	8

M	H	M	M	H	M
M	**M**	**H**	H	M	H
H	M	M	H	H	H
H	M	H	H	M	H
M	**M**	**H**	M	H	H
M	H	M	M	H	M
H	H	H	H	H	M
M	**M**	**H**	M	M	M

If a ballplayer has a 0.35 probability of hitting a fast ball, what is the probability of 2 misses before he gets a hit? Going from left to right and from the top row down, select the numbers in pairs. Let the pairs of integers from 01 to 35 in the above table represent a hit (H) by the ballplayer, and all the other pairs are a miss (M). The result is the table to the right. Out of the 16 triplet sets of letters, there are 3 sequences of MMH. Three of 16 is 19%. The analytically calculated value is 15%,
$g\ (z, p) = (1-p)^{z-1}p = (1-0.35)^{3-1}(0.35) = 0.148$, or 15%.

Expected value (mean or average) and variance for a binomial, geometric, and negative binomial distribution

The binomial distribution function,
$$b(k, n, p) = \frac{n!}{k!\,(n-k)!}p^k(1-p)^{n-k},$$
has two fixed parameters, n and p, and one variable k. The process associated with a binomial distribution may be thought of as n independent trials each with probability p of a

success, and k is the number of successes. The expected value, variance and standard deviation are, $E(z) = np$, $Var(z) = np(1-p)$, $std\ dev = \pm\sqrt{np(1-p)}$.

The geometric distribution function, $g(z,p) = (1-p)^{z-1}p$, has one fixed parameter p and one variable z. The process associated with a geometric distribution may be thought of as independent trials each with probability p of a success, and z is the number of trials until the first success. The expected value, variance and standard deviation are,

$$E(z) = \frac{1}{p},\ Var(z) = \frac{(1-p)}{p^2},\ std\ dev = \pm\sqrt{\frac{(1-p)}{p^2}}.$$

The negative binomial distribution function,

$$b_{neg}(w,r,p) = \frac{(w+r-1)!}{(r-1)!\,w!}p^r(1-p)^w,$$

has two fixed parameters r and p, and one variable w. The process associated with a negative binomial distribution may be thought of as independent trials each with probability p of success, and w is the number of failures before r successes. The expected value, variance and standard deviation are,

$$E(z) = \frac{r(1-p)}{p},\ Var(z) = \frac{r(1-p)}{p^2},\ std\ dev = \pm\sqrt{\frac{r(1-p)}{p^2}}.$$

Three equivalent statements of independence of two events A and B

If A and B are each a single event, they are defined to be independent by two equivalent statements:

The probability of A occurring is not conditional on event B occurring. This is written,
P(A|B)=P(A).
P(B|A)=P(B).

The probability of the intersection of A and B is the probability of A times the probability of B. This is written, P(A∩B)=P(A)P(B).

The statements are equivalent due to the definition of conditional probability,

$$P(B|A) = \frac{P(B \cap A)}{P(A)}\ \text{(eqn 1), and } P(A|B) = \frac{P(A \cap B)}{P(B)}\ \text{(eqn 2).}$$

Since intersection commutes, $B \cap A = A \cap B$, therefore, $P(B \cap A) = P(A \cap B)$, and $P(B|A)P(A) = P(A|B)P(B)$.

If P(A|B)=P(A) then P(B|A)=P(B), and by eqn 2 $P(A)P(B) = P(A \cap B)$, and
If P(B|A)=P(B) then P(A|B)=P(A).

Expectation (average) and variance of the sum and difference of two independent discrete random variables with arbitrary distributions and finite populations

From the definition of the expectation (average) of a discrete random variable with arbitrary distribution and a finite population,

$$E(X) = \sum_{i=1}^{i=N} x_i \, p_{i,}$$

where the population is size N and the probability of x_i is p_i.

From the definition of the variance of a discrete random variable with arbitrary distribution and a finite population,

$$Var(X) = \sum_{i=1}^{i=N} (x_i - E(X))^2 \, p_i.$$

For the sum or difference of two variables X and Y from different populations

$$E(X \pm Y) = \sum_{i=1}^{i=N_X} x_i \, p_{Xi} \pm \sum_{i=1}^{i=N_Y} x_i \, p_{Yi},$$

and

$$Var(X \pm Y) = \sum_{i=1}^{i=N_X} \left(x_i - E(X)\right)^2 p_{Xi} + \sum_{i=1}^{i=N_Y} (y_i - E(Y))^2 \, p_{Yi}.$$

Note: The expectation of the sum or difference of two variables retains the sign, but the variance of the sum or difference of two variables is always a positive summation.

Expectation (average) and variance of the sum and difference of two independent normally distributed variables

From the definition of the expectation (average) of a normally distributed variable,

$$E(X) = \int_{-\infty}^{+\infty} x f(x) dx = \mu,$$

where $f(x)$ is the probability distribution function (pdf) for the normal distribution $N(\mu, \sigma^2)$ with average μ and variance σ^2.

From the definition of the variance of a normally distributed variable,

$$Var(X) = \int_{-\infty}^{+\infty} (x - E(X))^2 f(x) dx = \sigma^2,$$

For the sum or difference of two independent normally distributed variables X and Y from different populations,

$$E(X \pm Y) = \int_{-\infty}^{+\infty} x f(x) dx \pm \int_{-\infty}^{+\infty} y g(y) dy = \mu_x \pm \mu_y,$$

where $f(x)$ is the pdf for $N(\mu_x, \sigma_x^2)$ and $g(x)$ is the pdf for $N(\mu_y, \sigma_y^2)$, and

$$Var(X \pm Y) = \int_{-\infty}^{+\infty} (x - E(X))^2 f(x) dx + \int_{-\infty}^{+\infty} (y - E(Y))^2 g(y) dy = \sigma_x^2 + \sigma_y^2.$$

Note: The expectation of the sum or difference of two variables retains the sign, but the variance of the sum or difference of two variables is always a positive summation.

Enumerate the properties of the normal distribution and the standard normal distribution

A normal distribution is represented by $\mathrm{N}(x, \mu, \sigma^2)$. The two fixed parameters are μ and σ^2, and x is the variable. However, as a practical matter, it is most useful to consider the standard normal distribution N(z,0,1) because data from any normal distribution can be normalized to distribute as N(z,0,1) by transforming the random variable X to

$$z = \frac{X - \mu}{\sigma} \sim N(z, 0,1).$$

The transformed variable z has a normal distribution with average 0 and variance 1. For $N(z, 0,1)$ the fixed parameters are $\mu = 0$ and $\sigma^2 = 1$, and z is the variable.

The distance from the midpoint of a normal distribution to the inflection point to the left or right is one standard deviation σ. For a standard normal distribution this corresponds to $z = \frac{\sigma}{\sigma} = 1$. The area under $\mathrm{N}(x, \mu, \sigma^2)$ from $-\sigma$ to $+\sigma$ is 68%, from -2σ to $+2\sigma$ is 95%, and -3σ to $+3\sigma$ is 99.7%. The area under N(z,0,1) from – 1 to +1 is 68%, from – 2 to +2 is 95%, and – 3 to +3 is 99.7%.

Relationship between the *z*-score, cdf and pdf for the standard normal distribution table

A z-score indicates the number of standard deviations a value is from the mean..

The Z-score calculation requires the values for the mean μ, the standard deviation σ and the particular value x of the random variable, using the formula $z = \frac{(x-\mu)}{\sigma}$. After determining the z-score, you may use the z-table to determine the probability that a z-score is less than a particular score, more than a particular score, or between two scores. For example, if you obtain a z-score of 2.91, you may wish to determine the probability that a score is less than that particular z-score, i.e., you want to determine the area under the curve that corresponds to that particular z-score. The table shows approximately 50%; thus, the probability that a z-score is less than 2.91 is roughly 50%.

The Standard Normal curve itself is the pdf (probability density function) of the standard normal distribution.

Each value in the table is the area under the curve to the left of the z value, and this area is obtained from the cdf (cumulative distribution function) evaluated for the value of z in the table. The cdf is the area under the curve of the pdf; this area is the same as integrating the pdf from $z = -\infty$ to the value of z in the table. Note. The area under the pdf is always 1.

Using the standard normal table when the illustration shows the tail

We are given that $X \sim N(\mu, \sigma^2)$. Suppose the population mean μ is 15 and a given x-value is 20. Using a significance level of $\alpha = 0.05$ what is the standard deviation σ?

Case #1: There is a picture of the pdf with the right tail shaded and labeled α.

Because $X - \mu > 0$, the tail is on the right. The table records the area to the right, so the tabulated area of $\alpha = 0.050$.
$z_\alpha = z_{0.05} = 1.645$, and
$$1.645 = \frac{(X - \mu)}{\sigma} = \frac{(20 - 15)}{\sigma} \rightarrow \sigma = 3.040$$

Note: If $0.50 < \alpha < 1.00$, you will not find z_α in this table. However, because of symmetry, $z_\alpha = -z_{1-\alpha}$, and $z_{1-\alpha}$ will be in this table.

Case #2: There is a picture of the pdf with the left tail shaded.

Because $X - \mu > 0$, the tail is on the right. The table records the area to the left, so look up the tabulated area for $(1 - \alpha) = 0.95$.
$z_{1-\alpha} = z_{0.95} = 1.645$, and
$$1.645 = \frac{(X - \mu)}{\sigma} = \frac{(20 - 15)}{\sigma} \rightarrow \sigma = 3.040$$

The t-Table only tabulates t for tails of area $0 < \alpha < 0.50$. The illustration above the table indicates that the tail is on the right. How can you find t when $0.50 < \alpha < 1.00$?

Because the t-Distribution is symmetrical about $t = 0$, $t_{1-\alpha} = -t_\alpha$.

Example: You want the value of $t_\alpha[n - 1] = t_{0.05}[10]$. Because $0 < 0.05 < 0.50$, you can use the table. The tabulated value of $t_{0.05}[10]$ is 1.812.

Example: You want the value of $t_{1-\alpha}[n - 1] = t_{0.95}[10]$. Because $0.50 < 0.95$, you will not find $t_{0.95}[n - 1]$. However, $t_{0.95}[n - 1] = -t_{0.05}[n - 1]$, so $t_{0.95}[10] = -t_{0.05}[10] = -1.812$. Such problems do not occur for the Chi-Square Table or the F-Table because the values for χ_α^2 and F_α are always positive.

Example:
$$\left(\frac{X - \mu}{\sigma}\right)^2 \sim\chi^2[df = 1]$$, where df is the degrees of freedom.

$X = 95, \mu = 100, \sigma = 5, \alpha = 0.05$. Find $\chi_\alpha^2[1]$.
The degree of freedom is 1. The degrees of freedom are listed in the column on the left of the Chi-Square Table.

$$\chi_\alpha^2[1] = \left(\frac{X - \mu}{\sigma}\right)^2 = \left(\frac{90.2 - 100}{5}\right)^2 = 3.842$$

The degree of freedom is 1. The degrees of freedom are listed in the column on the left of the Chi-Square Table.

Technique of using the standard normal table to estimate binomial statistics with large samples

Calculating binomial statistics with large samples involves the summation of many terms. It is often convenient to estimate binomial statistics using the standard normal table when np

and $n(1-p)$ exceed about 10. A more accurate estimate can be obtained if the binomial variable k has a continuity correction of $1/2$ made to it,
$b(k,n,p) \approx N\left(\left(k+\frac{1}{2}\right),\mu,\sigma^2\right) = N\left(\left(k+\frac{1}{2}\right),np,np(1-p)\right)$ when $np, np(1-p) \geq 10$.

Then the standard normal table can be used where
$$z = \frac{X-\mu}{\sigma} = \frac{(k+1/2)-np}{\sqrt{np(1-p)}}.$$

Recall that for the binomial distribution, $\mu = np$ and $\sigma^2 = np(1-p)$.

Example: If 70% of the Arizona population supports their new State law supporting enforcement of existing Federal immigration law, what is the probability that in a survey of 1000 people at most 700 people support the State law? 705 people? 750 people?

We are given that $p = 0.70, n = 1000$, and $k = 700$. So,
$$z = \frac{(k+1/2)-np}{\sqrt{np(1-p)}} = \frac{(700+1/2)-1000(0.70)}{\sqrt{1000(0.70)(1-0.70)}} = 0.0345,$$

The probability of at most 700 people? $z = 0.0345$ which corresponds to a probability of 0.51.

The probability of at most 705 people? $z = 0.380$ which corresponds to a probability of 0.648

The probability of at most 750 people? $z = 3.48$ which corresponds to a probability of 0.997.

The probability that between $\mu - \sigma$ and $\mu + \sigma$ support the State Law? $k = np \pm \sqrt{np(1-p)}$,
$$z = \frac{\left(np \pm \sqrt{np(1-p)} + 1/2\right) - np}{\sqrt{np(1-p)}} = \frac{\pm 14.4914 + 1/2}{14.4914} \cong \pm 1$$
This corresponds to a probability of 0.6826, the probability of being within $\pm\sigma$ of μ.

Constructing a Normal Probability Plot to determine if a random variable has a normal distribution

Order the data from smallest to largest.

Number the ordered data with the cardinal numbers $i = 1,2,3,\dots,n$.
Calculate $\frac{(i-0.5)}{n}$ for each ordered data point.

Treat each calculated value in Step 3 as the cumulative distribution function evaluated at z (the area to the left of z under the pdf for $N(0,1)$). Record the z value for each.

Plot the pairs (ordered value, z value) with the ordered value on the vertical axis and the z value on the horizontal axis, and pass a best fit line through the points

A fairly straight line indicates the data probably come from a normal distribution. The value on the vertical axis for $z = 0$ and $z = 1$ are the mean and standard deviation, respectively.

Show that for any arbitrary distribution, if $E(x) = \mu$ and $Var(x) = \sigma^2$, then $E(\bar{x}) = \mu$ and $Var(\bar{x}) = \frac{\sigma^2}{n}$.

If $E(x) = \mu$ and $Var(x) = \sigma^2$ for an arbitrary distribution, then, recalling that E(ax)=aE(x),

$$E(\bar{x}) = E\left(\frac{1}{n}\sum_{i=1}^{i=n} x_i\right) = \frac{1}{n}E\left(\sum_{i=1}^{i=n} x_i\right) = \frac{1}{n}\sum_{i=1}^{i=n} E(x_i) = \frac{1}{n}(n\mu) = \mu,$$

and, recalling that $Var(ax) = a^2Var(x)$ and $Var(x \pm y) = Var(x) + Var(y)$,

$$Var(\bar{x}) = Var\left(\frac{1}{n}\sum_{i=1}^{i=n} x_i\right) = \frac{1}{n^2}Var\left(\sum_{i=1}^{i=n} x_i\right) = \frac{1}{n^2}\left(\sum_{i=1}^{i=n} Var(x_i)\right) = \frac{1}{n^2}(n\sigma^2) = \frac{\sigma^2}{n}.$$

Note: As the sample size increases, the variance of the average goes to zero,

$$\lim_{n\to\infty} Var(\bar{x}) = \lim_{n\to\infty} \frac{\sigma^2}{n} = 0.$$

Example: For a binomial distribution, $E(k) = np$ and $Var(k) = np(1-p)$, so

$$E(\bar{k}) = E\left(\frac{k}{n}\right) = \frac{1}{n}E(k) = \frac{1}{n}np = p,$$

and

$$Var(\bar{k}) = Var\left(\frac{k}{n}\right) = \frac{1}{n^2}Var(k) = \frac{1}{n^2}np(1-p) = \frac{p(1-p)}{n}, \quad \lim_{n\to\infty} Var(\bar{k}) = 0.$$

Central Limit Theorem using the binomial distribution as an example

The second fundamental theorem of probability is the Central Limit Theorem (CLT). It states that as the sample size n of a variable X increases, the distribution of the sample average $\bar{x}$ approaches a normal distribution with mean μ (viz., $E(\bar{x}) = \mu$)and variance $\frac{\sigma^2}{n}$ (viz., $Var(\bar{x}) = \frac{\sigma^2}{n}$) irrespective of the distribution of the variable X.

Note: The X comes from an unknown distribution, but $\bar{x}_n$ "converges in distribution" as $n \to \infty$ to a normal distribution. "Converging in distribution" does <u>not</u> mean $\bar{x}_n$ converges to a specific value; it means the <u>distribution</u> of $\bar{x}_n$ converges to a normal distribution. This is written,

$\bar{X}_n$ converges in distribution as $\to \infty$ to $N\left(\mu, \frac{\sigma^2}{n}\right)$.

This can also be written,

$\frac{\bar{X}_n - \mu}{\sigma/\sqrt{n}}$ converges in distribution as $n \to \infty$ to $N(1,0)$.

Example: For a binomial distribution, $E(k) = np$ and $Var(k) = np(1-p)$, so

$$E(\bar{k}) = E\left(\frac{k}{n}\right) = \frac{1}{n}E(k) = \frac{1}{n}np = p,$$

and

$$Var(\bar{k}) = Var\left(\frac{k}{n}\right) = \frac{1}{n^2}Var(k) = \frac{1}{n^2}np(1-p) = \frac{p(1-p)}{n},$$

$$\frac{\frac{k}{n} - p}{\sqrt{\frac{p(1-p)}{n}}} \text{ converges in distribution as } n \to \infty \text{ to } N(1,0).$$

Show that if X is a Normally distributed random variable with $E(X) = \mu$ and $Var(X) = \sigma^2$,

$$Z = \frac{\bar{X}_n - \mu}{\sqrt{\frac{\sigma^2}{n}}} \sim N(0,1).$$

Proof: Recalling that $E(a) = a$, where a is a constant, $E(X \pm Y) = E(X) \pm E(Y)$ and

$$E(\bar{x}) = E\left(\frac{1}{n}\sum_{i=1}^{i=n} x_i\right) = \frac{1}{n}\sum_{i=1}^{i=n} E(x_i) = \frac{1}{n}n\mu = \mu, \text{ then}$$

$$E\left(\frac{\bar{X}_n - \mu}{\sqrt{\frac{\sigma^2}{n}}}\right) = \frac{1}{\sqrt{\frac{\sigma^2}{n}}}E(\bar{X}_n - \mu) = \frac{1}{\sqrt{\frac{\sigma^2}{n}}}\{E(\bar{X}_n) - E(\mu)\} = \frac{1}{\sqrt{\frac{\sigma^2}{n}}}\{\mu - \mu\} = 0$$

Recalling that $Var(aX) = a^2Var(X)$, $Var(X \pm Y) = Var(X) + Var(Y)$, and $Var(a) = 0$, where a is a constant,

$$Var\left(\frac{\bar{X}_n - \mu}{\sqrt{\frac{\sigma^2}{n}}}\right) = \frac{1}{\frac{\sigma^2}{n}}Var(\bar{X}_n - \mu) = \frac{n}{\sigma^2}\{Var(\bar{X}_n) + Var(\mu)\} = \frac{n}{\sigma^2}\left\{\frac{\sigma^2}{n} + 0\right\} = 1$$

Given two independent sample proportions k_1/n_1 and k_2/n_2 which are estimates $\hat{p}_1$ and $\hat{p}_2$ of probabilities p_1 and p_2 for the binomial distributions $b(k_1, n_1, p_1)$ and $b(k_1, n_1, p_1)$, find $E(p_1 - p_2)$ and $Var(p_1 - p_2)$ and show that even though k_1/n_1 and k_2/n_2 are not normally distributed,

$$\frac{\left(\frac{k_1}{n_1} - \frac{k_2}{n_2}\right) - (p_1 - p_2)}{\sqrt{\frac{p_1(1-p_1)}{n_1} + \frac{p_2(1-p_2)}{n_2}}} \textbf{ converges in distribution as } n_1, n_2 \to \infty \textbf{ to } N(0,1).$$

If the true population proportions are not known, $\frac{\left(\frac{k_1}{n_1} - \frac{k_2}{n_2}\right) - (p_1 - p_2)}{\sqrt{\frac{p_1(1-p_1)}{n_1} + \frac{p_2(1-p_2)}{n_2}}}$ converges in distribution to a normal distribution as $n_1, n_2 \to \infty$ according to the Central Limit Theorem.

It has average zero and variance 1 in the limit,

$$E\left(\frac{\left(\frac{k_1}{n_1}-\frac{k_2}{n_2}\right)-(p_1-p_2)}{\sqrt{\frac{p_1(1-p_1)}{n_1}+\frac{p_2(1-p_2)}{n_2}}}\right)=\frac{E\left(\frac{k_1}{n_1}-\frac{k_2}{n_2}-(p_1-p_2)\right)}{\sqrt{\frac{p_1(1-p_1)}{n_1}+\frac{p_2(1-p_2)}{n_2}}}=\frac{((p_1-p_2)-(p_1-p_2))}{\sqrt{\frac{p_1(1-p_1)}{n_1}+\frac{p_2(1-p_2)}{n_2}}}$$
$$=0$$

$$Var\left(\frac{\left(\frac{k_1}{n_1}-\frac{k_2}{n_2}\right)-(p_1-p_2)}{\sqrt{\frac{p_1(1-p_1)}{n_1}+\frac{p_2(1-p_2)}{n_2}}}\right)=\frac{Var\left(\left(\frac{k_1}{n_1}-\frac{k_2}{n_2}\right)-(p_1-p_2)\right)}{\frac{p_1(1-p_1)}{n_1}+\frac{p_2(1-p_2)}{n_2}}$$
$$=\frac{\frac{p_1(1-p_1)}{n_1}+\frac{p_2(1-p_2)}{n_2}-0}{\frac{p_1(1-p_1)}{n_1}+\frac{p_2(1-p_2)}{n_2}}=1$$

Therefore, the Z-Table (Standard Normal Table) may be used if n_1 and n_2 are large enough.

$n \geq Max\left(z_{\alpha/2}/[0.21\hat{p}], z_{\alpha/2}/[0.21(1\text{-}\hat{p})]\right)$. For a 95% confidence level, after rounding up to nearest integer, $n\hat{p}, n(1-\hat{p}) \geq 1.96/0.21 = 9.33 \rightarrow 10$.

For $X \sim N(\mu, \sigma^2)$ show that

$$\frac{\bar{X}_n-\mu}{\sqrt{\frac{\sigma^2}{n}}} \sim \text{N}(0,1)$$

Given a sample mean $\bar{X}_n$ of a Normally distributed random variable X with $E(X) = \mu$ and $Var(X) = \sigma^2$, determine its distribution.

Since $E(\bar{x}) = \mu$, and $Var(\bar{x}) = \frac{\sigma^2}{n}$, then $\bar{X}_n \sim N\left(\mu, \frac{\sigma^2}{n}\right)$.

Since it is more convenient to use the standard normal tables of $N(0,1)$, we write

$$\frac{\bar{X}_n-\mu}{\sqrt{\frac{\sigma^2}{n}}} \sim \text{N}(0,1),$$

and when estimating variance with s^2 for a sufficiently large n,

$$\frac{\bar{X}_n-\mu}{\sqrt{\frac{s^2}{n}}} \sim \text{N}(0,1).$$

For the Z-Table (Standard Normal Distribution) to give the same result within 2 significant figures as the T-Table using estimates of the variance depends on the sample size and confidence level: $n \geq 20$ at 90% C.L., $n \geq 30$ at 95% C.L., $n \geq 80$ at 99% C.L.

If $X \sim N\left(\mu_x, \frac{\sigma_x^2}{n_x}\right)$ and $Y \sim N\left(\mu_y, \frac{\sigma_y^2}{n_y}\right)$
show that for sufficiently large n_x and n_y

$$\frac{(\bar{X}_{nx} - \bar{Y}_{ny}) - (\mu_x - \mu_y)}{\sqrt{s_x^2/n_x + s_y^2/n_y}} \sim \mathrm{N}(0,1),$$

If $X \sim N\left(\mu_x, \frac{\sigma_x^2}{n_x}\right)$ and $Y \sim N\left(\mu_y, \frac{\sigma_y^2}{n_y}\right)$ show that

$$\frac{(\bar{X}_{nx} - \bar{Y}_{ny}) - (\mu_x - \mu_y)}{\sqrt{\sigma_x^2/n_x + \sigma_y^2/n_y}} \sim \mathrm{N}(0,1),$$

and for sufficiently large n_x and n_y show that

$$\frac{(\bar{X}_{nx} - \bar{Y}_{ny}) - (\mu_x - \mu_y)}{\sqrt{s_x^2/n_x + s_y^2/n_y}} \sim \mathrm{N}(0,1),$$

Proof:

$$E\left(\frac{(\bar{X}_{nx} - \bar{Y}_{ny}) - (\mu_x - \mu_y)}{\sqrt{s_x^2/n_x + s_y^2/n_y}}\right) = \frac{1}{\sqrt{\sigma_x^2/n_x + \sigma_y^2/n_y}} E\left((\bar{X}_{nx} - \bar{Y}_{ny}) - (\mu_x - \mu_y)\right) = 0$$

$$Var\left(\frac{(\bar{X}_{nx} - \bar{Y}_{ny}) - (\mu_x - \mu_y)}{\sqrt{s_x^2/n_x + s_y^2/n_y}}\right) = \frac{1}{\sigma_x^2/n_x + \sigma_y^2/n_y} Var\left((\bar{X}_{nx} - \bar{Y}_{ny}) - 0\right)$$

$$= \frac{1}{\sigma_x^2/n_x + \sigma_y^2/n_y}\left(\frac{\sigma_x^2}{n_x} + \frac{\sigma_y^2}{n_y}\right) = 1$$

For the Z-Table (Standard Normal Distribution) to give the same result within 2 significant figures as the T-Table using estimates of the variance depends on the sample size and confidence level: $n \geq 20$ at 90% C.L., $n \geq 30$ at 95% C.L., $n \geq 80$ at 99% C.L.

General form for a variable that has a Standard Normal Distribution, $N(0,1)$..Some of the useful variables that distribute as $N(0,1)$

The general form of a variable that has a Standard Normal Distribution $N(0,1)$ is, $Z \sim N(0,1)$, where Z is an independent and identically distributed (iid) variable.

$$\frac{X - \mu}{\sigma} \sim N(0,1),$$

$$\frac{\bar{X}_n - \mu}{\sigma/\sqrt{n}} \sim N(0,1),$$

$$\frac{(\bar{x} - \bar{y}) - (\mu_x - \mu_y)}{\sqrt{\sigma_x^2/n_x + \sigma_y^2/n_y}} \sim N(0,1),$$

$$\frac{(\bar{x}-\bar{y})-(\mu_x-\mu_y)}{\sqrt{s_x^2/n_x+s_y^2/n_y}} \sim N(0,1), \text{ when } (n_x+n_y-2)>30,$$

$$\frac{\frac{k}{n}-p}{\sqrt{\frac{p(1-p)}{n}}} \sim N(1,0),$$

$$\frac{\left(\frac{k_1}{n_1}-\frac{k_2}{n_2}\right)-(p_1-p_2)}{\sqrt{\frac{p_1(1-p_1)}{n_1}+\frac{p_2(1-p_2)}{n_2}}} \sim N(0,1).$$

Chi-Square Distribution

The symbol for the Chi-Square Distribution is $\chi^2(v)$ where v is the degrees of freedom. Just as $N(0,1)$ is the symbol for a standard normal distribution but is not the actual pdf, so $\chi^2(v)$ is the symbol for the Chi-Square Distribution but is not the actual pdf.

The pdf of the Chi-Square Distribution, $f(X,v)$, is a complicated function of a variable X and a constant parameter v, the degrees of freedom.

If an independent and identically distributed (iid) variable $X \sim N(0,1)$, then

$$\left(\frac{X-\mu}{\sigma}\right)^2 \sim \chi^2[1],$$

$$\sum_{i=1}^{i=n}\left(\frac{X_i-\mu}{\sigma}\right)^2 \sim \chi^2[n],$$

$$\frac{(n-1)s^2}{\sigma^2} \sim \chi^2[n-1].$$

Note: The symbol s^2 refers to the estimate of σ^2.

A chi-square test is often used to compare percentages to a mean percentage to determine goodness of fit, i.e., significance of the model. In other words, it is a comparison of observed frequencies to the expected value (mean).

The Chi-Square Distribution is written $\chi^2(v)$, where v is the degrees of freedom. In the Chi-Square Table the degrees of freedom are listed in the left column.

The general form of a variable that has a Chi-Square Distribution is,

$$\sum_{i=1}^{i=v} Z_i^2 \sim \chi^2(v), \text{ where } Z_i \sim N(0,1) \text{ for all } i.$$

Note: See the list of variables that distribute as $N(0,1)$.

Specific variables that have a Chi-Square -Distribution are,

1. $\left(\frac{X-\mu}{\sigma}\right)^2 \sim \chi^2[1],$

$$2.\sum_{i=1}^{i=n}\left(\frac{X_i-\mu}{\sigma}\right)^2 \sim \chi^2[n],$$

$$3.\frac{(n-1)s^2}{\sigma^2} \sim \chi^2[n-1],$$

$$4.\frac{\left(\frac{k}{n}-p\right)^2}{\frac{p(1-p)}{n}} \sim \chi^2[1].$$

Using the Chi-Square table

A portion of the Chi-Square table is below. A picture of the distribution is usually on top.

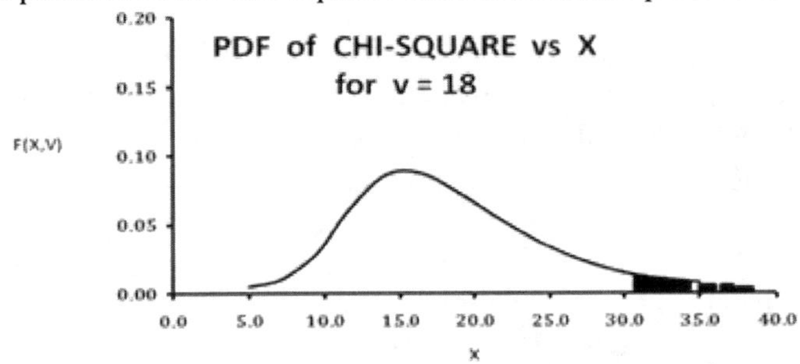

Y (DEGREES OF FREEDOM)	AREA OF TAIL						
	0.975	0.950	.	0.050	0.025	0.010	0.005
1	<<0.0001	<<0.001	.	3.84	5.02	6.63	7.88
2	0.051	0.103	.	5.99	7.38	9.21	10.6
3	0.216	0.352	.	7.81	9.35	11.3	12.8
.	.	.	.	.	.	.	.
.	.	.	.	.	.	.	.
18	8.23	9.39	.	28.9	31.5	34.8	37.2

Example 1:
$X = \chi^2_{0.025}[18] = 31.5$

Example 2:
$X = \chi^2_{0.975}[18] = 8.23$Therefore, in order for a model to be significant at the .025 level, with degrees of freedom of 18, the chi squared value would need to be 31.5 or lower. In order for a model to be significant at the 0.975 level (an unlikely significance choice), the chi squared value would need to be 8.23 or lower. Normally, you have calculated the chi squared value or have used a statistical software package that computes the chi squared value. Once you have the value, you can manually check for significance, using the table. For example, suppose you find the chi-squared value to be 15.6 for 8 degrees of freedom. Find the closest value in the chart for the particular degrees of freedom given. Doing so reveals a p-value of .05, which is significant.

Variable that has a T-Distribution

The general form for the expressions that have a T-Distribution is,

$$\frac{U}{\sqrt{V/v}} \sim t(v),$$

where U is any variable that has a $N(0,1)$ distribution and
V is any variable that has a $\chi^2(v)$ distribution. Also, U and V must be independent variables.
Example:

$$U = \frac{\bar{X}_n - \mu}{\sigma/\sqrt{n}} \sim N(0,1),$$

$$V = \frac{(n-1)s^2}{\sigma^2} \sim \chi^2[n-1],$$

$$\frac{U}{\sqrt{V/v}} = \frac{\frac{\bar{X}_n - \mu}{\sigma/\sqrt{n}}}{\sqrt{\frac{(n-1)s^2}{\sigma^2(n-1)}}} = \frac{\bar{X}_n - \mu}{s/\sqrt{n}}$$

$$\frac{\bar{X}_n - \mu}{s/\sqrt{n}} \sim t[n-1].$$

Useful variables that have a T-Distribution

The general form for the expressions that have a T-Distribution is,

$$\frac{U}{\sqrt{V/v}} \sim t[v],$$

$U \sim N(0,1), V \sim \chi^2[v]$, U and V must be independent variables.

Specific variables that have a T-Distribution are,

1. $\frac{\bar{X}_n - \mu}{s/\sqrt{n}} \sim t[n-1],$

2. $\frac{(\bar{X} - \bar{Y}) - (\mu_X - \mu_Y)}{\sqrt{\left(\frac{(n_X - 1)s_X^2 + (n_Y - 1)s_Y^2}{n_X + n_Y - 2}\right)}\sqrt{\left(\frac{1}{n_X} + \frac{1}{n_Y}\right)}} \sim t[n_X + n_Y - 2],$

3. $\frac{\bar{D} - \mu_D}{s_{\bar{D}}} \sim t[n-1],$
   where $X \sim N(\mu_X, \sigma_X^2)$, $Y \sim N(\mu_Y, \sigma_Y^2)$, $\mu_D = E(X_i - Y_i)$,

4. $\bar{D}_n = \frac{1}{n}\sum_{i=1}^{i=n}(X_i - Y_i)$, and $s_{\bar{D}}^2 = \frac{\sum_{i=1}^{i=n}(D_i^2) - n\bar{D}^2}{n-1}$.

Standard error of the mean

The standard error of the mean (SEM) is the square root of the variance of the mean estimated from a sample of the population. It is typically used when the true variance of the population is not known.

For a binomially distributed variable $k \sim b(k, n, p)$ the mean and variance of the mean are,

$$E(\bar{k}) = E\left(\frac{k}{n}\right) = \frac{1}{n}E(k) = \frac{1}{n}np = p,$$
$$Var(\bar{k}) = Var\left(\frac{k}{n}\right) = \frac{1}{n^2}Var(k) = \frac{1}{n^2}np(1-p) = \frac{p(1-p)}{n}.$$
The estimate of p is $\hat{p} = \bar{k}_n = {}^{k}/_{n}$, and the SEM (square root of the variance of the mean) is
$$SEM(\bar{k}_n) = SEM(\hat{p}) = \sqrt{\frac{\hat{p}(1-\hat{p})}{n}}.$$
For a normally distributed variable $x \sim N(0,1)$ the mean and variance of the mean are,
$E(\bar{x}) = \mu$ and $Var(\bar{x}) = \frac{\sigma^2}{n}$.
The estimate of the mean and estimate of the variance of the mean are,
$$\bar{x}_n = \frac{1}{n}\sum_{i=1}^{i=n} x_i, \text{ and } \frac{\hat{\sigma}^2}{n} = \frac{s^2}{n} = \frac{1}{n}\left\{\frac{1}{n-1}\sum_{i=1}^{i=n}(x_i - \bar{x}_n)^2\right\}.$$

The SEM (square root of the variance of the mean) is
$$SEM(\bar{x}_n) = \frac{s}{\sqrt{n}}.$$

Populations of continuous, discrete and categorical variables, and the distributions that may be associated with each

Some populations consist of continuous variables and may be associated with normal distribution, uniform continuous distributions, or other continuous distributions. Such populations, even when they are not associated with any distribution or the distribution is unknown, will nearly always have a population mean and variance.

Some populations consist of discrete variables and are associated with discrete distributions such as the uniform discrete distribution, the binomial distribution, or the multinomial distribution. Each outcome is identified with a discrete numerical value.

Some populations consist of categorical variables and are associated with binomial, multinomial, geometric, negative binomial, or hypergeometric distributions. Variables may be assigned values from sample sets containing two outcomes, such as {success, failure}, {red ball, white ball} or {yes, no} and are associated with a binomial distribution. Variables may be assigned values from sample sets with more than two outcomes, such as {yes for candidate A, yes for candidate B, yes for candidate C} and are associated with a multinomial distribution.

Inferential Statistics and Correlation and Regression

Classifications of estimates of population parameters

Estimates of population parameters may be classified into two groups, point estimates and interval estimates.

A point estimate would be the estimate of $E(x) = E(\bar{x}) = \mu$ from $\bar{x}_n = \frac{1}{n}\sum_{i=1}^{i=n} x_i$, the estimate of $Var(x) = \sigma^2$ from $s^2 = \frac{1}{n-1}\sum_{i=1}^{i=n}(x_i - \bar{x}_n)^2$, or the estimate of $Var(\bar{x}) = \sigma^2/n$ from s^2/n.

An interval estimate is the upper and lower bound of a population parameter. Often the distribution of the variable is known. For example, given $\bar{x}_n$ where X is normally distributed,

it will also be true that

$$\frac{\bar{X}_n - \mu}{s_n/\sqrt{n}} \sim t(n-1).$$

So, for two different values of t ($-A$ and A) on the horizontal axis of the T-Distribution, we can get an upper and lower bound on the value of μ,

$$Pr(-A < t < A) = 0.90,$$

$$Pr\left(-A < \frac{\bar{X}_n - \mu}{s_n/\sqrt{n}} < A\right) = 0.90 \text{ and}$$

The interval estimate for μ is,

$$\bar{x}_n \pm A\frac{s_n}{\sqrt{n}} \text{ with a confidence level of } 90\%$$

Using the Chebyshev inequality

When the distribution of a variable X is unknown, but the population average μ and population variance σ^2 are known, one may use Chebyshev's inequality to estimate the probability that the value of x will be $\pm k\sigma$ from μ,

$$Pr(\mu - k\sigma \leq X \leq \mu + k\sigma) \geq \left(1 - \frac{1}{k^2}\right).$$

Example: Suppose that a manufacturer has assembled 10,000 units of a metal rod whose population mean length is 150 mm and population standard deviation is 10 mm. With no knowledge of what the distribution looks like, we can let $k = 2$ and state that at least $(1 - \frac{1}{2^2})$ or $\frac{3}{4}$ of the rods are within ± 2 standard deviations of the mean. Using the above notation, $Pr(\mu - 2\sigma \leq X \leq \mu + 2\sigma) \geq \frac{3}{4}$.

The length is 150 mm ± 20 mm with a confidence level of 75%.

Method of making an interval estimate when the distribution of the variable is unknown and the average and variance are also unknown

When the distribution of a variable X is unknown, and the population average and population variance are also unknown, one can take the following steps to determine if the variable has a known distribution. We use the standard normal distribution in the following example:

Order the data from smallest to largest.

Number the ordered data with the cardinal numbers $i = 1,2,3, \dots, n$.

Calculate $\frac{(i-0.5)}{n}$ for each ordered data point.

Treat each calculated value in Step 3 as the cumulative distribution function evaluated at z (the area to the left of z under the pdf for $N(0,1)$). Record the z value for each.

Plot the pairs (ordered value, z value) with the ordered value on the vertical axis and the z value on the horizontal axis, and pass a best fit line through the points.

A fairly straight line indicates the data probably come from a normal distribution, and we may proceed with the next step.
Calculate the point estimates for the population mean and average.
$Pr(\bar{x}_n - s_n \leq \mu \leq \bar{x}_n + s_n) \cong 0.6826$, and
$Pr(\bar{x}_n - 1.96s_n \leq \mu \leq \bar{x}_n + 1.96s_n) \cong 0.95$.

The interval estimate for μ is,
$\bar{x}_n \pm 1.96s_n$ with a confidence level of 95%.

Accuracy and precision

The margin of error is a measure of precision. Precision is defined as the "repeatability" of a measurement. It is distinguished from the accuracy. The accuracy is a measure of the correct or true value. For example, the illustrations below compare precision with accuracy using a target or bulls eye.

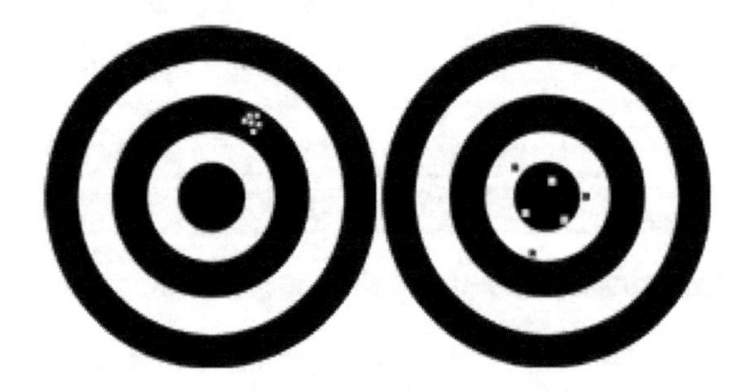

The target on the left demonstrates precision. The bullet holes are very close together, however, not in the center.

The target on the right demonstrates accuracy, because all the bullet holes are in the center circle. They lack the precision of the target on the left because the bullet holes are farther apart, but they are more accurately placed.

Precision can be determined from a sample of the population. According to the law of large numbers, the precision of the calculated average will improve as the sample size increases. The calculated variance of $\bar{x}$ will go to zero as the sample size increases, according to the formula s^2/n. Accuracy of the mean and variance can only be determined with a knowledge of the true population mean and variance. Accuracy requires the absence of bias. That includes sample bias as well as bias in the estimator.

Example: A chemist conducts an experiment to measure the content of an active ingredient in a drug. His result is precise, that is to say, repeatable. However, the analytical method gives a lower than true result because of the presence of another chemical. Because of this interference, his method does not produce accurate results. The FDA requires that analytical methods in the pharmaceutical industry must have both precision and accuracy.

Population parameter estimate

A sampling method is a survey or an experimental procedure that is said to be verified if it is both precise and accurate. Determining precision is relatively simple, using point estimators.

Proving accuracy can be more difficult.

There are two basic methods to verify accuracy. One involves multiple sampling methods in a survey or multiple experimental procedures. For example, in a pharmaceutical analysis of a drug, the result from a chromatographic technique can be compared to that from a spectrophotometric technique and a gravimetric technique.

The other method involves increasing the true population mean by a known amount.

Example: Workers take samples to measure the average length of a machined part. The manager can test their accuracy by adding a number of machined parts of known length. The relative increase in the average should be close to the fraction of parts that are added by the manager times the relative increase in length of the added parts.

How the standard error of $\hat{p}$, $SE(\hat{p})$, in a sample survey is estimated using a binomial or multinomial distribution

The standard error of the mean (SEM) is a point estimate, and it is an estimate of precision. It will measure the random error in a sampling survey, but it will not measure the bias. The distribution in a survey of a population may be binomial if there are only two possible outcomes, or it may be multinomial if there are more than two possible outcomes. A multinomial distribution is a generalization of the binomial distribution. For a multinomial distribution, the average and variance for the ith outcome in a total of τ outcomes is,
$\mathrm{E}(k_i) = np_i, Var(k_i) = np_i(1 - p_i)$, where $\sum_{i=1}^{i=\tau} p_i = 1$.
The estimate $\hat{p}_i$ for each p_i is $\hat{p}_i = k_i/n$.

The estimates for the means and variances are,

$$\frac{k_i}{n_i} = \hat{p}_i,\ SEM(\hat{p}_i) = \sqrt{\frac{\hat{p}_i(1-\hat{p}_i)}{n_i}}.$$

Example: A survey of 1000 people for the top three choices for senatorial candidate in the Democratic primary gave the following results, for candidates A, B and C, $k_A = 300, k_B = 600$, and $k_C = 100$. $\hat{p}_A = 0.300, \hat{p}_B = 0.600, \hat{p}_C = 0.100$.

$$SEM_A = \sqrt{\frac{0.3(1-0.3)}{300}} = 0.026, SEM_B = \sqrt{\frac{0.6(1-0.6)}{600}} = 0.020, \text{ and } SEM_C = \sqrt{\frac{0.1(1-0.1)}{100}} = 0.0300.$$

Candidate A: 30.0%±2.6%, Candidate B: 60.0%±2.0%, Candidate C: 10.0%±3.0%.

Notations for the population parameter, its estimate, the standard deviation and the standard error

Θ is the population parameter

$\hat{\Theta}$ is the estimate of the population parameter

$$Std\ Dev(\Theta) = \sqrt{Var(\Theta)}$$
$$SE(\hat{\Theta}) = \sqrt{Estimate\ of\ Var(\Theta)}$$

Example:

$\Theta = \mu, \hat{\Theta} = \bar{x}_n,$

$Std\ Dev(\Theta) = \sigma/\sqrt{n}$

$$SE(\hat{\Theta}) = s_n/\sqrt{n} = \left(\frac{1}{n-1}\sum_{i=1}^{i=n} x_i\right)/\sqrt{n}$$

$$Pr\left((\bar{x}_n - Z_{\alpha/2}\,\sigma/\sqrt{n}) < \mu < (\bar{x}_n + Z_{\alpha/2}\,\sigma/\sqrt{n})\right) = (1-\alpha)100\%.$$

$$Pr\left((\bar{x}_n - t_{\alpha/2}[n-1]\,s_n/\sqrt{n}) < \mu < (\bar{x}_n + t_{\alpha/2}[n-1]\,s_n/\sqrt{n})\right) = (1-\alpha)100\%.$$

Example:

$\Theta = p, \hat{\Theta} = \hat{p} = k/n$

$$Std\ Dev(\Theta) = \sqrt{\frac{p(1-p)}{n}}$$

$$SE(\hat{\Theta}) = \sqrt{\frac{\hat{p}(1-\hat{p})}{n}} = \sqrt{\frac{k/n(1-k/n)}{n}}$$

$$Pr\left(\left(p - Z_{\alpha/2}\sqrt{\frac{k/n(1-k/n)}{n}}\right) < \mu < \left(p + Z_{\alpha/2}\sqrt{\frac{k/n(1-k/n)}{n}}\right)\right) = (1-\alpha)100\%.$$

How large should n be for one to use the Z-Table (Standard Normal Table) instead of the T-Table?

Note: The entries in a Z-Table are tail areas, the z-values are in the margin; the entries in a T-Table are the t values, the tail areas are in the top row.

To determine the size of n required to get the same result within 2 significant figures as the Z-Table, observe the entries below, taken from a T-Table,

Y (DEGREES OF FREEDOM)	AREA OF TAIL (α/2)		
	0.050	0.025	0.005
10	1.812	2.228	3.169
20	1.725	2.086	2.845
30	1.697	2.042	2.750
40	1.684	2.021	2.704
50	1.676	2.009	2.678
60	1.671	2.000	2.660
80	1.664	1.990	2.639
100	1.660	1.984	2.626
1000	1.646	1.962	2.581
∞	1.645	1.960	2.576
	90%	95%	99%
	CONFIDENCE LEVEL		

Note: The degrees of freedom for the t-Distribution is $[n - 1]$.

For 90%, 95% and 99% confidence levels, z is (to 2 significant figures) 1.7, 2.0, and 2.6.

For 90% t is 1.7 for $n = 20$; for 95% t is 2.0 for $n = 30$; for 99% t is 2.6 for $n = 80$.

For the Z-Table to give the same result within 2 significant figures as the T-Table depends on the confidence level as well as the sample size:

$n \geq 20$ at 90% C.L., $n \geq 30$ at 95% C.L., $n \geq 80$ at 99% C.L.

Properties of point estimators

Unbiasedness: An estimator, $\widehat{\Theta}$, is an equation that estimates a population parameter Θ.

$\widehat{\Theta}$ is an unbiased estimator if $E(\widehat{\Theta}) - \Theta = 0$.

Consistency: The estimates of the estimator, $\widehat{\Theta}$, get closer to the population parameter, Θ, as the sample size increases, and the values of the estimates are distributed evenly on both sides of Θ.

Precision: The size of the variance of the estimator $E\left[\left(\widehat{\Theta} - \mathrm{E}(\Theta)\right)^2\right]$ is a measure of the precision of the estimator. The smaller the variance, the more precise the estimator.

Example:
The variance of the median of a normal distribution is $(\pi/_{2n})\sigma^2$.

The variance of the mean of a normal distribution is $\sigma^2/_n$.

So, the variance of the mean is more precise.

Unbiasedness

Bias can be defined as estimator bias. An estimator $\widehat{\Theta}$ is an equation that estimates a population parameter Θ.

$\widehat{\Theta}$ is an unbiased estimator if $E(\widehat{\Theta}) - \Theta = 0$.
$\widehat{\Theta}$ is a biased estimator with $Bias(\widehat{\Theta})$, if $E(\widehat{\Theta}) - \Theta = Bias(\widehat{\Theta}) \neq 0$.

Example: Given $X \sim N(\mu, \sigma^2)$, then let $\widehat{\Theta} = \hat{\mu}$ and $\Theta = \mu$.

$\hat{\mu} = \frac{\sum_{i=1}^{i=n} x_i}{n}$ is an unbiased estimator of μ, because

$$E(\hat{\mu}) - \mu = E\left(\frac{\sum_{i=1}^{i=n} x_i}{n}\right) - \mu = \frac{1}{n}E\left(\sum_{i=1}^{i=n} x_i\right) - \mu = \frac{1}{n}\sum_{i=1}^{i=n} E(x_i) - \mu = \frac{n}{n}\mu - \mu = 0.$$

Example: Given $X \sim N(\mu, \sigma^2)$, then let $\widehat{\Theta} = \widehat{\sigma_1^2}$ and $\Theta = \sigma^2$.

$$= \frac{\sum_{i=1}^{i=n}(x_i - \bar{x}_n)^2}{n}$$

is a biased estimator of σ^2, because

$$E\left(\widehat{\sigma_1^2}\right) - \sigma^2 = E\left(\frac{\sum_{i=1}^{i=n}(x_i - \bar{x}_n)^2}{n}\right) - \sigma^2 = \frac{n-1}{n}\sigma^2 - \sigma^2 = -\frac{\sigma^2}{n}.$$

Therefore, $\frac{\sum_{i=1}^{i=n} x_i^2 - n\bar{x}^2}{n}$

is a biased estimator of σ^2 with bias = $-\frac{\sigma^2}{n}$, and

$$\widehat{\sigma_1^2} = \frac{\sum_{i=1}^{i=n}(x_i - \bar{x}_n)^2}{n-1}$$

is an unbiased estimator of σ^2.

Variance of the mean and the median of a normal distribution

The mean, typically the arithmetic average, minimizes the expected value of the squared deviation. This is written as,

$E[(X - X_{mean})^2] \leq E[(X - A)^2]$, for any value A of the variable X.

The median is the value for which half the data is smaller and half the data is larger, and it minimizes the expected value of the absolute deviation. This is written as,
$E[|X - X_{median}|] \leq E[|(X - A)|]$, for any value A of the variable X.

The variance of the median of a normal distribution is $({}^{\pi}/_{2n})\sigma^2$.

The variance of the mean of a normal distribution is ${}^{\sigma^2}/_{n}$. In summary,
$X \sim N(\mu, \sigma^2)$,
$X_{median,\,n} \sim N(\mu, ({}^{\pi}/_{2n})\sigma^2))$,

$$\bar{X}_n \sim N(\mu, {}^{\sigma^2}/_{n}.$$

Confidence level

A confidence level of say 95% means that if the same sampling method were repeated a total of 100 times and the confidence interval was calculated each time, the true population parameter would fall inside 95 of the calculated confidence intervals.

Example: A normally distributed variable X is sampled 10 times. The standard deviation and 95% confidence interval is calculated. This process is repeated 10 times. For a normal distribution, 95% of the values are between $\mu - 1.96\sigma$ and $\mu + 1.96\sigma$. The results for the 10 estimates of the 95% confidence interval are,

#	$\bar{x}_{10} - 1.96s_n$	$\bar{x}_{10} + 1.96s_n$
1	94	116
2	94	112
3	95	119
4	94	110
5	91	109
6	91	107
7	95	121
8	100	114
9	94	110
10	95	107

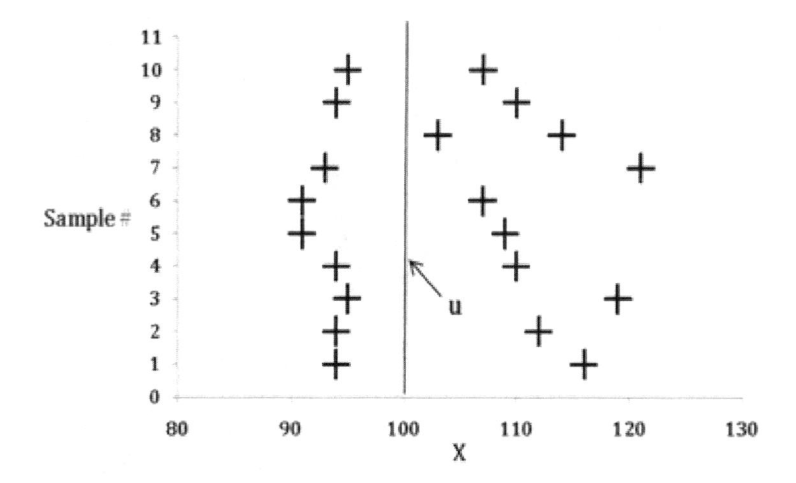

It turns out that $\mu = 100$, therefore, the true population mean does not fall in the confidence interval for sample #8, but it does fall within 9 of the 10 confidence intervals which is reasonably close to the 95% confidence level.

Confidence interval

A confidence interval consists of three components: a statistic, a margin of error, and a confidence level.

The format is $sample\ statistic \pm margin\ of\ error\ with\ confidence\ level\ of\ XX\%$.
The margin of error is the square root of the estimate of the variance, and it is calculated from sample data.

When the distribution of the variable is known, the confidence level may be calculated by the area under the probability distribution function (pdf) and between the upper and lower limits of the confidence interval of the variable. This area is listed in the table for the distribution function, for example the standard normal table.

There is no single way of constructing the confidence interval of an unobservable population parameter, Θ. It depends on the statistical variable whose interval is to be estimated and the distribution of that statistical variable (if any exist). The first rule of constructing an interval is that the upper and lower end-points, U and L, are both functions of the statistical variable X. In other words, the confidence interval will be $(L(X), U(X))$. The probability Pr that the confidence interval $(L(X), U(X))$ includes the unobservable population parameter Θ at the $(1-\alpha)(100\%)$ confidence level is,

$$Pr\big(L(X) < \Theta < U(X)\big) = (1-\alpha).$$

Constructing confidence interval using T-Distribution

Example: If $X \sim N(\mu, \sigma^2)$, the two unobservable population parameters are the mean μ and the variance σ^2. The estimate of the variance is $s_n^2 = 25, n = 11$. We first observe that

$$\frac{\bar{X}_n - \mu}{s_n/\sqrt{n}} \sim t[n-1], \text{ and}$$

$$Pr\left(L\big(t_{\alpha/2}[n-1], \bar{X}_n\big) < \mu < U\big(t_{\alpha/2}[n-1], \bar{X}_n\big)\right) = (1-\alpha) = 0.95.$$

$$Pr\left(-t_{\alpha/2}[n-1] < \frac{\bar{X}_n - \mu}{s_n/\sqrt{n}}, \frac{\bar{X}_n - \mu}{s_n/\sqrt{n}} < t_{\alpha/2}[n-1]\right)$$

$$= Pr\left(t_{\alpha/2}[n-1] > \frac{-\bar{X}_n + \mu}{s_n/\sqrt{n}}, \frac{-\bar{X}_n + \mu}{s_n/\sqrt{n}} > -t_{\alpha/2}[n-1]\right)$$

$$= Pr\left(t_{\alpha/2}[n-1]\frac{s_n}{\sqrt{n}} > -\bar{X}_n + \mu, -\bar{X}_n + \mu > -t_{\alpha/2}[n-1]\frac{s_n}{\sqrt{n}}\right)$$

$$= Pr\left(\bar{X}_n + t_{\alpha/2}[n-1]\frac{s_n}{\sqrt{n}} > \mu, \mu > \bar{X}_n - t_{\alpha/2}[n-1]\frac{s_n}{\sqrt{n}}\right)$$

$$= Pr\left(\bar{X}_n - t_{0.05/2}[11-1]\frac{5}{\sqrt{11}} < \mu < \bar{X}_n + t_{0.05/2}[11-1]\frac{5}{\sqrt{11}}\right)$$

$$= Pr\left(-2.228\frac{5}{\sqrt{11}} < \mu < \bar{X}_n + 2.228\frac{5}{\sqrt{11}}\right) = (1-\alpha) = 0.95$$

$$(-3.359, 3.359)$$

Standard cost analysis

If the T-Distribution is used, the confidence interval is

$$\left(-t_{\alpha/2}[n-1]\frac{s_n}{\sqrt{n}}, t_{\alpha/2}[n-1]\frac{s_n}{\sqrt{n}}\right), \text{ and}$$

$$\bar{X}_n \pm t_{\alpha/2}\frac{s_n}{\sqrt{n}} \text{ with confidence level } (1-a)100\%.$$

If the Z-Distribution is used, the confidence interval is

$$\left(-z_{\alpha/2}\frac{\sigma}{\sqrt{n}}, z_{\alpha/2}\frac{\sigma}{\sqrt{n}}\right), \text{ and}$$

$$\bar{X}_n \pm z_{/2}\frac{\sigma}{\sqrt{n}} \text{ with confidence level } (1-a)100\%.$$

For probabilities using binomial distributions where sample size is sufficiently high,

$$\left(-z_{\alpha/2}\sqrt{\frac{\hat{p}(1-\hat{p})}{n}}, z_{\alpha/2}\sqrt{\frac{\hat{p}(1-\hat{p})}{n}}\right), \text{ and}$$

$$\hat{p} \pm z_{\alpha/2}\sqrt{\frac{\hat{p}(1-\hat{p})}{n}}$$ with confidence level $(1-a)100\%$.

As the confidence levels decreases, the margin of error increases. This is because as a decreases, both $t_{\alpha/2}$ and $z_{\alpha/2}$ decrease.

As the sample size increases, the margin of error decreases and the confidence level increases. This is because the margin of error has the sample size in its denominator.

As the population standard deviation decreases, the margin of error decreases. This is because the standard deviation is in the numerator of the expression for the margin of error.

Derive the expression for z in terms of the binomial distribution so that the z-Table may be used to construct a confidence interval $(-z_{1-\alpha/2}, z_{1-\alpha/2})$ for a proportion p at the $(1-\alpha)100\%$ level

A proportion on a sample space S={A,B} implies a binomial distribution or multinomial distribution if S={A,B,C...}. For a binomial distribution $b(k, n, p)$, outcome A occurs k times, each time with probability p, and outcome B occurs $(n-k)$ times, each time with probability $(1-p)$. If this sample of n outcomes is repeated numerous times, and number of values for k are observed which reveal that k has a binomial distribution (viz., $k \sim b(k, n, p)$).

Recall that for a binomial distribution
$E(k) = np, Var(k) = E[(k - E(k))^2] = np(1-p)$, and
$E\left(\frac{k}{n}\right) = p, Var\left(\frac{k}{n}\right) = \frac{1}{n^2}Var(k) = \frac{1}{n^2}\{np(1-p)\} = \frac{p(1-p)}{n}$.
By the Central Limit Theorem, for any X that has any arbitrary distribution,
$$z = \frac{X - E(X)}{\sqrt{Var(X)}}$$
will converge in distribution, as the sample size goes to infinity, to $N(1,0)$. For $X = k$,
$$z = \frac{k - np}{\sqrt{np(1-p)}} = \frac{\frac{k}{n} - p}{\sqrt{\frac{p(1-p)}{n}}} = \frac{\hat{p} - p}{\sqrt{\frac{p(1-p)}{n}}}$$
converges in distribution as $n \to \infty$ to $N(0,1)$.
Also, when $Var(X)$ is replaced with its unbiased estimator
$\frac{\hat{p}(1-\hat{p})}{n}$,
$$z = \frac{\hat{p} - p}{\sqrt{\frac{\hat{p}(1-\hat{p})}{n}}}$$
Also converges in distribution as $n \to \infty$ to $N(0,1)$.

Using the expression for z in terms of the binomial distribution, construct a confidence interval $(-z_{\alpha/2}, z_{\alpha/2})$ for p at the $(1-\alpha)100\%$ level for sufficiently large n

$$z = \frac{\hat{p} - p}{\sqrt{\frac{\hat{p}(1-\hat{p})}{n}}}$$

lies in the interval $(-z_{\alpha/2}, z_{\alpha/2})$ at the $(1-\alpha)100\%$ level for sufficiently large n.

$$Pr\left(-z_{\alpha/2} < \frac{\hat{p} - p}{\sqrt{\frac{\hat{p}(1-\hat{p})}{n}}}, \frac{\hat{p} - p}{\sqrt{\frac{\hat{p}(1-\hat{p})}{n}}} < z_{\alpha/2}\right) = (1-\alpha)$$

$$= Pr\left(z_{\alpha/2} > \frac{-\hat{p} + p}{\sqrt{\frac{\hat{p}(1-\hat{p})}{n}}}, \frac{-\hat{p} + p}{\sqrt{\frac{\hat{p}(1-\hat{p})}{n}}} > -z_{\alpha/2}\right)$$

$$= Pr\left(\hat{p} + z_{\alpha/2}\sqrt{\frac{\hat{p}(1-\hat{p})}{n}} > p, p > \hat{p} - z_{\alpha/2}\sqrt{\frac{\hat{p}(1-\hat{p})}{n}}\right)$$

$$= Pr\left(\hat{p} + z_{\alpha/2}\sqrt{\frac{\hat{p}(1-\hat{p})}{n}} > p > \hat{p} - z_{\alpha/2}\sqrt{\frac{\hat{p}(1-\hat{p})}{n}}\right) = (1-\alpha).$$

$$\left(-z_{\alpha/2}\sqrt{\frac{\hat{p}(1-\hat{p})}{n}}, z_{\alpha/2}\sqrt{\frac{\hat{p}(1-\hat{p})}{n}}\right)$$

How large must the sample size be to calculate the confidence interval with no more than 10% error?

When sample size is too small, the confidence interval at the $(1-\alpha)100\%$ confidence level

$$\hat{p} \pm z_{\alpha/2}\sqrt{\frac{\hat{p}(1-\hat{p})}{n}}$$

must be replaced with

$$\frac{\hat{p} + \frac{1}{2n}z_{\alpha/2}^2 \pm z_{\alpha/2}\sqrt{\frac{\hat{p}(1-\hat{p})}{n} + \frac{1}{4n^2}z_{\alpha/2}^2}}{1 + \frac{1}{n}z_{\alpha/2}^2}.$$

So, for the $z_{\alpha/2}^2$ terms to contribute less than 10% error in the interval for a $(1-\alpha)$ confidence level,

$$n \geq Max\left(\frac{z_{\alpha/2}}{0.21\hat{p}}, \frac{z_{\alpha/2}}{0.21(1-\hat{p})}\right).$$

Note: For a 95% confidence level, after rounding up to nearest integer,

$$n\hat{p}, n(1-\hat{p}) \geq \frac{1.96}{0.21} = 9.33 \rightarrow 10.$$

Example: What is the minimum sample size required to use the z-Table to calculate the 95% confidence interval so that there is no more than 10% error in the interval when $\hat{p} = 0.10$?

$$n \geq Max\left(\frac{z_{\alpha/2}}{0.21\hat{p}}, \frac{z_{\alpha/2}}{0.21(1-\hat{p})}\right) = Max\left(\frac{z_{0.025}}{0.21(0.10)}, \frac{z_{0.025}}{0.21(0.90)}\right)$$
$$= Max\left(\frac{1.96}{0.21(0.10)}, \frac{1.96}{0.21(0.90)}\right) = Max\left(\frac{1.96}{0.21(0.10)}, \frac{1.96}{0.21(0.90)}\right) = Max(93.3, 10.4) = 94.$$

Note: More samples are needed when $\hat{p}$ is very low or very high.

Derive the expression for z in terms of the binomial distribution so that the z-Table may be used to construct a confidence interval $(-z_{\alpha/2}, z_{\alpha/2})$ for the difference in two proportions $(p_1 - p_2)$ at the $(1-\alpha)100\%$ level. Next, construct a confidence interval $(-z_{\alpha/2}, z_{\alpha/2})$

By the Central Limit Theorem, if k_1 and k_2 each have binomial distributions $b_1(k_1, n_1, p_1)$ and $b_2(k_2, n_2, p_2)$, and the $Unbiased\ Estimator\ of\ Mean = (\hat{p}_1 - \hat{p}_2)$, and the $Unbiased\ Estimator\ of\ Var(p_1 - p_2) = \hat{p}_1(1-\hat{p}_1)/n_1 + \hat{p}_2(1-\hat{p}_2)/n_2$, then
$$z = \frac{(\hat{p}_1 - \hat{p}_2) - (p_1 - p_2)}{\sqrt{\hat{p}_1(1-\hat{p}_1)/n_1 + \hat{p}_2(1-\hat{p}_2)/n_2}}$$
converges in distribution as $n \to \infty$ to $N(0,1)$. For sufficiently large n, we construct the probability that z lies in the interval $(-z_{\alpha/2}, z_{\alpha/2})$ at the $(1-\alpha)100\%$ level.

$$Pr\left(-z_{\alpha/2} < \frac{(\hat{p}_1 - \hat{p}_2) - (p_1 - p_2)}{\sqrt{\frac{\hat{p}_1(1-\hat{p}_1)}{n_1} + \frac{\hat{p}_2(1-\hat{p}_2)}{n_2}}}, \frac{(\hat{p}_1 - \hat{p}_2) - (p_1 - p_2)}{\sqrt{\frac{\hat{p}_1(1-\hat{p}_1)}{n_1} + \frac{\hat{p}_2(1-\hat{p}_2)}{n_2}}} < z_{\alpha/2}\right) = (1-\alpha)$$

$$= Pr\left((\hat{p}_1 - \hat{p}_2) + z_{\alpha/2}\sqrt{\frac{\hat{p}_1(1-\hat{p}_1)}{n_1} + \frac{\hat{p}_2(1-\hat{p}_2)}{n_2}}\right.$$
$$\left. > (p_1 - p_2) > (\hat{p}_1 - \hat{p}_2) - z_{\alpha/2}\sqrt{\frac{\hat{p}_1(1-\hat{p}_1)}{n_1} + \frac{\hat{p}_2(1-\hat{p}_2)}{n_2}}\right) = (1-\alpha).$$

$$\left(-z_{\alpha/2}\sqrt{\frac{\hat{p}_1(1-\hat{p}_1)}{n_1} + \frac{\hat{p}_2(1-\hat{p}_2)}{n_2}}, z_{\alpha/2}\sqrt{\frac{\hat{p}_1(1-\hat{p}_1)}{n_1} + \frac{\hat{p}_2(1-\hat{p}_2)}{n_2}}\right)$$

Derive the confidence interval at confidence level $(1-\alpha)100\%$ for the mean μ of a normally distributed variable X when the population variance σ^2 is known

For $X \sim N(\mu, \sigma^2)$,
$E(X) = \mu, Var(X) = \sigma^2$, and
$$z = \frac{X - E(X)}{\sqrt{Var(X)}} = \frac{X - \mu}{\sigma} \sim N(1,0).$$
Let $X = \bar{X}_n$, then
$$z = \frac{\bar{X}_n - E(\bar{X}_n)}{\sqrt{Var(\bar{X}_n)}} = \frac{\bar{X}_n - \mu}{\sqrt{\frac{\sigma^2}{n}}} = \frac{\bar{X}_n - \mu}{\sigma/\sqrt{n}}$$

We construct the probability that z lies in the interval $\left(-z_{\alpha/2}, z_{\alpha/2}\right)$ at the $(1-\alpha)100\%$ level.

$$Pr\left(-z_{\alpha/2} < \frac{\bar{X}_n - \mu}{\sigma/\sqrt{n}}, \frac{\bar{X}_n - \mu}{\sigma/\sqrt{n}} < z_{\alpha/2}\right) = (1-\alpha)$$

$$= Pr\left(\bar{X}_n + z_{\alpha/2}\frac{\sigma}{\sqrt{n}} > \mu > \bar{X}_n - z_{\alpha/2}\frac{\sigma}{\sqrt{n}}\right) = (1-\alpha).$$

The confidence interval for the population mean at the $(1-\alpha)100\%$ confidence level is,

$$\left(-z_{\alpha/2}\frac{\sigma}{\sqrt{n}}, z_{\alpha/2}\frac{\sigma}{\sqrt{n}}\right).$$

Example: A variable X has a population variance $\sigma^2 = 5$. The sample size is 25. Find the confidence interval for the 95% confidence level.

$$\left(-z_{\alpha/2}\frac{\sigma}{\sqrt{n}}, z_{\alpha/2}\frac{\sigma}{\sqrt{n}}\right)$$

$$= \left(-1.96\frac{5}{\sqrt{25}}, 1.96\frac{5}{\sqrt{25}}\right) = (-1.96, 1.96).$$

Derive the confidence interval at confidence level $(1-\alpha)100\%$ for the mean μ of a normally distributed variable X when the population variance σ^2 is not known

By the Central Limit Theorem, for $X \sim N(\mu, \sigma^2)$,

$$\frac{Unbiased\ Estimate\ of\ \bar{X} - E(\bar{X})}{\sqrt{Unbiased\ Estimate\ of\ Var(\bar{X})}} = \frac{\bar{x}_n - \mu}{\sqrt{s_n^2/n}} = \frac{\bar{x}_n - \mu}{s_n/\sqrt{n}} \sim t[n-1].$$

We construct the probability that μ lies in the interval $\left(-t_{\alpha/2}[n-1], t_{\alpha/2}[n-1]\right)$ at the $(1-\alpha)100\%$ level.

$$Pr\left(-t_{\alpha/2}[n-1] < \frac{\bar{X}_n - \mu}{s_n/\sqrt{n}}, \frac{\bar{X}_n - \mu}{s_n/\sqrt{n}} < t_{\alpha/2}[n-1]\right) = (1-\alpha)$$

$$= Pr\left(\bar{X}_n + t_{\alpha/2}[n-1]\frac{s_n}{\sqrt{n}} > \mu > \bar{X}_n - t_{\alpha/2}[n-1]\frac{s_n}{\sqrt{n}}\right) = (1-\alpha).$$

$$\left(-t_{\alpha/2}[n-1]\frac{s_n}{\sqrt{n}}, t_{\alpha/2}[n-1]\frac{s_n}{\sqrt{n}}\right)$$

Example: If $s_n = 5$, the sample size is 25, and the confidence level is 95%, what is the confidence interval?

$$\left(-t_{\alpha/2}[n-1]\frac{s_n}{\sqrt{n}}, t_{\alpha/2}[n-1]\frac{s_n}{\sqrt{n}}\right) = \left(-2.064\frac{5}{\sqrt{25}}, 2.064\frac{5}{\sqrt{25}}\right) = (-2.064, 2.064).$$

The difference procedure is determining the confidence interval when the population variance σ^2 is known and when it must be estimated from the data as s_n^2.

Example: If $\bar{X}_n = 20$, $n = 25$, and population variance σ^2 is 25, what is the confidence interval at the 95% confidence level?

$$Pr\left(\bar{X}_n + z_{\alpha/2}\frac{\sigma}{\sqrt{n}} > \mu > \bar{X}_n - z_{\alpha/2}\frac{\sigma}{\sqrt{n}}\right) = (1-\alpha)$$

$$= Pr\left(\bar{X}_n + z_{0.05/2}\frac{5}{\sqrt{25}} > \mu > \bar{X}_n - z_{0.05/2}\frac{5}{\sqrt{25}}\right) = (1-0.05)$$

$$= Pr\left(\bar{X}_n + 1.96\frac{5}{\sqrt{25}} > \mu > \bar{X}_n - 1.96\frac{5}{\sqrt{25}}\right) = 0.95$$

$$= Pr(\bar{X}_n + 1.96 > \mu > \bar{X}_n - 1.96) = 0.95.$$

$$\left(-z_{\alpha/2}\frac{\sigma}{\sqrt{n}}, z_{\alpha/2}\frac{\sigma}{\sqrt{n}}\right) = (-1.96, 1.96)$$

Example: If $\bar{X}_n = 20, n = 25$, and estimate of population variance s_n^2 is 25, what is the confidence interval at the 95% confidence level?

$$Pr\left(\bar{X}_n + t_{\alpha/2}[n-1]\frac{s_n}{\sqrt{n}} > \mu > \bar{X}_n - t_{\alpha/2}[n-1]\frac{s_n}{\sqrt{n}}\right) = (1-\alpha)$$

$$= Pr\left(\bar{X}_n + t_{0.05/2}[25-1]\frac{5}{\sqrt{25}} > \mu > \bar{X}_n - t_{0.05/2}[25-1]\frac{5}{\sqrt{25}}\right) = (1-0.05)$$

$$= Pr\left(\bar{X}_n + 2.064\frac{5}{\sqrt{25}} > \mu > \bar{X}_n - 2.064\frac{5}{\sqrt{25}}\right) = 0.95.$$

$$\left(-t_{\alpha/2}[n-1]\frac{s_n}{\sqrt{n}}, t_{\alpha/2}[n-1]\frac{s_n}{\sqrt{n}}\right) = (-2.064, 2.064)$$

Note: $t_{\alpha/2}[n-1]$ is a value on the x-axis of the $t[n-1]$ distribution, and $-t_{1-\alpha/2}[n-1] = t_{\alpha/2}[n-1]$ because $t[n-1]$ is symmetric about $t_{0.50}[n-1] = 0$.

Confidence interval of the variance at the $(1-\alpha)100\%$ confidence level

For $X \sim N(\mu, \sigma^2)$,

$$\frac{(n-1)s_n^2}{\sigma^2} \sim \chi^2[n-1]$$

Note: The entries in a Chi-Square Table are values of $\chi^2_{\alpha/2}[n-1]$ for the x-axis of the pdf curve, the degrees of freedom $[n-1]$ are in the left column, and the areas of the tail $\alpha/2$ are in the top row.

$$Pr\left(\chi^2_{1-\alpha/2}[n-1] < \frac{(n-1)s_n^2}{\sigma^2}, \frac{(n-1)s_n^2}{\sigma^2} < \chi^2_{\alpha/2}[n-1]\right) = (1-\alpha),$$

$$Pr\left(\sigma^2 < \frac{(n-1)s_n^2}{\chi^2_{1-\alpha/2}[n-1]}, \frac{(n-1)s_n^2}{\chi^2_{\alpha/2}[n-1]} < \sigma^2\right) = (1-\alpha),$$

$$Pr\left(\chi^2_{1-\alpha/2}[n-1] < \frac{(n-1)s_n^2}{\sigma^2}, \frac{(n-1)s_n^2}{\sigma^2} < \chi^2_{\alpha/2}[n-1]\right) = (1-\alpha),$$

$$Pr\left(\sigma^2 < \frac{(n-1)s_n^2}{\chi^2_{1-\alpha/2}[n-1]}, \frac{(n-1)s_n^2}{\chi^2_{\alpha/2}[n-1]} < \sigma^2\right) = (1-\alpha),$$

$$Pr\left(\frac{(n-1)s_n^2}{\chi^2_{\alpha/2}[n-1]} < \sigma^2 < \frac{(n-1)s_n^2}{\chi^2_{1-\alpha/2}[n-1]}\right) = (1-\alpha),$$

$$\left(\frac{(n-1)s_n^2}{\chi^2_{\alpha/2}[n-1]}, \frac{(n-1)s_n^2}{\chi^2_{1-\alpha/2}[n-1]}\right) \text{ at the } (1-\alpha)100\% \text{ confidence level.}$$

Note: $\chi^2_{1-\alpha/2}$ and $\chi^2_{\alpha/2}[n-1]$ are values on the x-axis of the $\chi^2[n-1]$ distribution, the values on the x-axis are always positive, and $\chi^2[n-1]$ is not symmetric.

Show that the confidence interval of the ratio of two variances at the $(1-\alpha)100\%$ confidence level is

$$\left(\frac{s_1^2/s_2^2}{F_{\alpha/2}[r_1,r_2]}, F_{\alpha/2}[r_2,r_1]s_1^2/s_2^2\right).$$

If two random variables U and V have a Chi-Square Distribution with degrees of freedom $r_1$and r_2, respectively, then their ratio has an $F[r_1,r_2]$-Distribution. This is written, if $U\sim\chi^2[r_1]$ and $V\sim\chi^2[r_2]$, then

$$\frac{U/r_1}{V/r_2}\sim F[r_1,r_2].$$

For $X_1\sim N(\mu_1,\sigma_1^2)$, and $X_2\sim N(\mu_2,\sigma_2^2)$,

$$U=\frac{r_1 s_1^2}{\sigma_1^2}\sim\chi^2[r_1], \text{ and } V=\frac{r_2 s_2^2}{\sigma_2^2}\sim\chi^2[r_2], \text{ so}$$

$$F_{\alpha/2}[r_1,r_2]=\frac{U/r_1}{V/r_2}=\frac{s_1^2/\sigma_1^2}{s_2^2/\sigma_2^2}\sim F[r_1,r_2].$$

The area under the pdf of $F[r_1,r_2]$ and to the right of $F_{\alpha/2}[r_1,r_2]$ on the x-axis is $\alpha/2$.

$$Pr\left(F_{1-\alpha/2}[r_1,r_2]<\frac{s_1^2/\sigma_1^2}{s_2^2/\sigma_2^2},\frac{s_1^2/\sigma_1^2}{s_2^2/\sigma_2^2}<F_{\alpha/2}[r_1,r_2]\right)=(1-\alpha),$$

$$Pr\left(\frac{\sigma_1^2}{\sigma_2^2}<\frac{s_1^2/s_2^2}{F_{1-\alpha/2}[r_1,r_2]},\frac{\sigma_2^2}{\sigma_1^2}<\frac{F_{\alpha/2}[r_1,r_2]}{s_1^2/s_2^2}\right)=(1-\alpha),$$

$$Pr\left(\frac{\sigma_1^2}{\sigma_2^2}<\frac{s_1^2/s_2^2}{F_{1-\alpha/2}[r_1,r_2]},\frac{s_1^2/s_2^2}{F_{\alpha/2}[r_1,r_2]}<\frac{\sigma_1^2}{\sigma_2^2}\right)=(1-\alpha),$$

$$Pr\left(\frac{s_1^2/s_2^2}{F_{\alpha/2}[r_1,r_2]}<\frac{\sigma_1^2}{\sigma_2^2}<\frac{s_1^2/s_2^2}{F_{1-\alpha/2}[r_1,r_2]}\right)=(1-\alpha), \textit{but } F_{1-\alpha/2}[r_1,r_2]=\left(F_{\alpha/2}[r_2,r_1]\right)^{-1}, \text{so,}$$

$$\left(\frac{s_1^2/s_2^2}{F_{\alpha/2}[r_1,r_2]}, F_{\alpha/2}[r_2,r_1]s_1^2/s_2^2\right).$$

The confidence interval for the difference between two unpaired means? How large must the sample size be at the 90%, 95% and 99% confidence level to use the Z-Table?

For $X_1\sim N(\mu_1,\sigma_1^2), X_2\sim N(\mu_2,\sigma_2^2)$

$$z=\frac{(\bar{x}_1-\bar{x}_2)-E(\bar{x}_1-\bar{x}_2)}{\sqrt{Var(\bar{x}_1-\bar{x}_2)}}=\frac{(\bar{x}_1-\bar{x}_2)-(\mu_1-\mu_2)}{\sqrt{\sigma_1^2/n_1+\sigma_2^2/n_2}}\sim N(1,0).$$

We construct the probability that z lies in the interval $\left(-z_{\alpha/2},z_{\alpha/2}\right)$ at the $(1-\alpha)100\%$ level.

$$Pr\left(-z_{\alpha/2}<\frac{(\bar{x}_1-\bar{x}_2)-(\mu_1-\mu_2)}{\sqrt{\sigma_1^2/n_1+\sigma_2^2/n_2}},\frac{(\bar{x}_1-\bar{x}_2)-(\mu_1-\mu_2)}{\sqrt{\sigma_1^2/n_1+\sigma_2^2/n_2}}<z_{\alpha/2}\right)=(1-\alpha)$$

$$=Pr\left((\bar{x}_1-\bar{x}_2)+z_{\alpha/2}\sqrt{\sigma_1^2/n_1+\sigma_2^2/n_2}>(\mu_1-\mu_2)>(\bar{x}_1-\bar{x}_2)-z_{\alpha/2}\sqrt{\sigma_1^2/n_1+\sigma_2^2/n_2}\right).$$

The confidence interval for the population mean at the $(1-\alpha)100\%$ confidence level is,

$$\left(-z_{\alpha/2}\sqrt{\sigma_1^2/n_1+\sigma_2^2/n_2}, z_{\alpha/2}\sqrt{\sigma_1^2/n_1+\sigma_2^2/n_2}\right).$$

If the population variance is not known, one may estimate the variance from the data if n is large enough. Recall that for the Z-Table to give the same result within 2 significant figures as the T-Table using estimates of the variance depends on the sample size and confidence level:

$n \geq 20$ at 90% C.L., $n \geq 30$ at 95% C.L., $n \geq 80$ at 99% C.L.

$$\left(-z_{\alpha/2}\sqrt{s_1^2/n_1+s_2^2/n_2}, z_{\alpha/2}\sqrt{s_1^2/n_1+s_2^2/n_2}\right).$$

Note: $z_{\alpha/2}$ is a value on the x-axis of the $N(0,1)$ distribution (Standard Normal), and $-z_{1-\alpha/2} = z_{\alpha/2}$ because $N(0,1)$ is symmetric about $z_{0.50} = 0$.

The confidence interval for a difference between two paired means? Explain why paired means X and $Y$are dependent on each other?

Given $X \sim N(\mu_x, \sigma_x^2)$ and $Y \sim N(\mu_y, \sigma_y^2)$. Two paired means X and $Y$are dependent on each other because the act of pairing makes event x_i conditional to event y_i and vice-versa. However, this does not necessarily mean that X and $Y$are correlated.

Define:

$\mu_D = \mu_x - \mu_y$

$d_i = \mu_{xi} - \mu_{yi}$,

$\bar{d}_n = \frac{1}{n}\sum_{i-1}^{i=n}(\mu_{xi} - \mu_{yi})$, and

$s_D = \frac{1}{(n-1)}\sum_{i=1}^{i=n} {d_i}^2$

Since μ_{xi} and μ_{yi} are random normally distributed variables, d_i is also a random normally distributed variable, so

$$\frac{\bar{d}_n - \mu_D}{s_D/\sqrt{n}} \sim t[n-1], \text{ and}$$

$$Pr\left(-t_{\alpha/2}[n-1] < \frac{\bar{d}_n - \mu_D}{s_D/\sqrt{n}}, \frac{\bar{d}_n - \mu_D}{s_D/\sqrt{n}} < t_{\alpha/2}[n-1]\right) = (1-\alpha)$$

$$= Pr\left(\bar{d}_{n_n} - t_{\alpha/2}[n-1]\frac{s_D}{\sqrt{n}} < \mu < \bar{d}_n + t_{\alpha/2}[n-1]\frac{s_D}{\sqrt{n}}\right) = (1-\alpha)$$

The confidence interval for the average difference μ_D at the $(1-\alpha)100\%$ confidence level is,

$$\left(-t_{\alpha/2}[n-1]\frac{s_D}{\sqrt{n}}, t_{\alpha/2}[n-1]\frac{s_D}{\sqrt{n}}\right)$$

Give an example of calculating the confidence interval at the 95% confidence level for a linear least-squares regression line using the SLOPE, STDEV and STEYX functions in an EXCEL spreadsheet

The set of data $\{(x_i, y_i)\}$ is in the left table. The terms and their values using the functions in EXCEL are in the right table.

xi	yi
10	8
11	10
12	8
13	16
14	13
15	19
16	15
17	30
18	20
19	28
20	24

FUNCTION IN EXCEL	TERM	VALUE
SLOPE	Slope	1.981818
INTERCEPT	Intercept	-12.3636
STEYX	SE(ycap)	4.089405
STDEV	SE(x)	3.316625

$$\widehat{Var}(Slope) = \frac{\widehat{Var}(\hat{y})}{(n-1)\widehat{Var}(x)},$$

$$SE(Slope) = \sqrt{\widehat{Var}(Slope)} = \frac{\sqrt{\widehat{Var}(\hat{y})}}{\sqrt{(n-1)}\sqrt{\widehat{Var}(x)}} = \frac{SE(\hat{y})}{\sqrt{(n-1)}SE(x)} = \frac{4.089405}{(\sqrt{10})3.316625}$$

$$= 0.390$$

$t_{\alpha/2}[n-2] = t_{0.025}[9] = 2.262,$

$Margin\ of\ Error\ of\ Slope = (t_{0.025}[9])SE(Slope) = (2.262)(0.390) = 0.882.$

$Slope \pm Margin\ of\ Error\ at\ 95\%\ Confidence\ Level =$

1.982 ± 0.882 at the 95% Confidence Level.

Population parameters and sample parameters

Population parameters are characteristics of the population, whereas sample statistics are characteristics of a sample of the population. The following are some characteristics of populations and characteristics of samples of those populations.

N	Number of elements in a finite population	n	Number of observations in sample from a finite or infinite population
N_i	Number of elements in the ith finite population	n_i	Number of observations in ith sample from a finite or infinite population
p	Proportion of outcome A_1 in population. Sample space is $\{A_1, A_2\}$.	$\hat{p}$	k/N, where k is number of outcome A_1in sample. Sample space is $\{A_1, A_2\}$.
p_i	Proportion of outcome A_1 in ith population. Sample space is $\{A_1, A_2\}$.	$\hat{p}_i$	k_i/N_i, where k_i is number of outcome A_1in sample from ith population. Sample space is $\{A_1, A_2\}$.

Population parameters are characteristics of the population, whereas sample statistics are characteristics of a sample of the population. The following are some characteristics of populations and characteristics of samples of those populations.

$p_{j,i}$	Proportion of outcome A_j in ith population. Sample space is $\{A_1,.., A_j..\}$.	$\hat{p}_{j,i}$	$k_{j,i}/N_i$, where $k_{j,i}$ is number of outcome A_j in ith population. Sample space is $\{A_1,.., A_j..\}$.
μ	Mean of a finite or infinite population	$\bar{x}_n$	Estimate of mean calculated from sample of size n. Subscript often omitted.
μ_i	Mean of the ith finite or infinite population	$\bar{x}_{n,i}$	Estimate of mean ith finite or infinite population, calculated from sample of size n. First subscript often omitted.
σ^2	Population variance of X	s_n^2	Estimate of variance of X calculated from sample of size n. Subscript often omitted.
σ_p^2	Population variance of p	SEM_p	Standard error of p, equal to $\sqrt{\hat{p}(1-\hat{p})/n}$
$\sigma_{\bar{X}}^2$	Population variance of $\bar{X}$ ($\bar{X} = \mu$)	$SEM_{\bar{X}}$	Standard error of $\bar{X}$, equal to $\sqrt{s_n^2/n}$

Significance testing

The concept of significance testing is to:

Propose two hypotheses, a null hypothesis and an alternate hypothesis, about the equality or inequality of two population parameters.

Transform a sample data set into a random variable with a presumed distribution.

Evaluate the area under the pdf of the distribution, from the value for the transformed variable (the critical value) out to the tail. Equate this area to the probability that the sample data set supports either the null hypothesis or alternate hypothesis.

The population parameters examined in this way are the mean, the variance or the proportion (μ, σ^2 or p, respectively).

A null hypothesis is assumed to be true and the user either rejects or fails to reject the null hypothesis, based upon the resulting significance, or p-value. For example, if the significance level, or alpha, is set by the researcher at .01, a p-value of .005 would cause the researcher to reject the null hypothesis and declare a significant difference between the groups. A p-value of .02 would cause the researcher to fail to reject the null hypothesis and thus declare no significant difference between the groups.

Hypotheses tested, transformed variables, and presumed distributions

Some of the hypotheses tested, the transformed variables, and the presumed distributions are:

H_0	H_1	TRANSFORMED VARIABLE	DISTRIBUTION
$\mu - \mu_0 = 0$	$\mu - \mu_0 > 0$	$z = \frac{\bar{x} - \mu_0}{\sigma/\sqrt{n}}$	$N(0,1)$
$\mu_x - \mu_y = 0$	$\mu_x - \mu_y > 0$	$z = \frac{\bar{x} - \bar{y} - 0}{\sqrt{\frac{\sigma_x^2}{n} + \frac{\sigma_y^2}{m}}}$	$N(0,1)$
$\mu_x - \mu_y = 0$	$\mu_x - \mu_y > 0$	$z = \frac{\bar{x} - \bar{y} - 0}{\sqrt{\frac{s_x^2}{n} + \frac{s_y^2}{m}}}, n \gg 0$	$N(0,1)$
$\mu - \mu_0 = 0$	$\mu - \mu_0 > 0$	$t[n-1] = \frac{\bar{x} - \mu_0}{s/\sqrt{n}}$	T-Distribution
$\sigma - \sigma_0^2 = 0$	$\sigma - \sigma_0^2 > 0$	$\chi^2[n-1] = \frac{(n-1)s^2}{\sigma_0^2}$	Chi-Square
$\frac{\sigma_x^2}{\sigma_y^2} - 1 = 0$	$\frac{\sigma_x^2}{\sigma_y^2} - 1 > 0$	$F[n-1, m-1] = \frac{s_x^2}{s_y^2}$	F-Distribution
$p - p_0 = 0$	$p - p_0 > 0$	$z = \frac{\frac{k}{n} - p_0}{\sqrt{p_0(1-p_0)/n}}$	$N(0,1)$

Null hypothesis

The null hypothesis H_0 generally claims an equality relationship between one or more population parameters. The form of a null hypothesis is a function $H(\theta_1, \theta_2) = 0$, where θ_1 and θ_2 are population parameters.

The null hypothesis $\mu_x - \mu_y = 0$ is equivalent to ${}^{\mu_x}/_{\mu_y} - 1 = 0$, but the latter is not used because it can be shown (using a Taylor expansion of ${}^{x}/_{y}$ about the point (μ_x, μ_y)) that ${}^{\bar{x}}/_{\bar{y}}$ is not an unbiased estimator of ${}^{\mu_x}/_{\mu_y}$; rather, one would have to use

$$\frac{\bar{x}}{\bar{y}} - \frac{1}{n\bar{y}^2}\left(s_y^2 \frac{\bar{x}}{\bar{y}} - s_{xy}\right).$$

The null hypothesis ${}^{\sigma_x^2}/_{\sigma_y^2} - 1 = 0$ is equivalent to $\sigma_x^2 - \sigma_y^2 = 0$, but the latter is not used.

The estimate of the former, ${}^{s_x^2}/_{s_y^2}$, can be transformed into a variable that has an F-distribution:

$$\frac{s_x^2}{s_y^2}\frac{\sigma_y^2}{\sigma_x^2} \sim F[r_x, r_y].$$

However, neither the estimate of the latter, $s_x^2 - s_y^2$, nor any transformation of it has a known distribution from which to calculate a probability.

Alternative hypothesis

While the null hypothesis H_0 generally claims an equality relationship between two population parameters, the alternative hypothesis H_1 generally claims an inequality relationship between the same two population parameters. The null hypothesis is the "hypothesis of no difference".

The form of a null hypothesis is a function, $H_0: H(\theta_1, \theta_2) = 0$ (hence, the name "null" hypothesis) where θ_1 and θ_2 are population parameters.

The form of an alternate hypothesis is the same function, $H_1: H(\theta_1, \theta_2) > 0$.

When the estimate of the alternative hypothesis is transformed into a variable $G(\hat{\theta}_1, \theta_2)$ with a known distribution, the sample data is used to evaluate the variable (the critical value). The probability of that value or a lesser value occurring may then be used to support the hypothesis.

P-value and how the different levels are interpreted regarding support of the null hypothesis: p <0.05, 0.05< p <0.10, and 0.10< p

When the estimate of a null hypothesis, H_0 is transformed into a variable with a known distribution, the sample data is used to evaluate the variable. The area under the pdf of the distribution from the evaluated variable (critical value) on the x-axis to the tail is the probability of that value or a lesser value occurring. This probability may then be used to support the null hypothesis. The probability is called the *p-value*. The smaller the *p-value*, the less support the sample data gives to the null hypothesis.

Often, the following guidelines are used:
When $p < 0.05$, it is considered that there is enough evidence from the sample data to reject the null hypothesis. This corresponds to a 5% chance of rejecting the null hypothesis when it is true (Type I error).

When $0.05 < p < 0.10$, it is considered that there is some evidence from the sample data against the null hypothesis.
When $p > 0.10$, it is considered that there is little or no evidence from the sample data to reject the null hypothesis.

Note. The researcher determines the sufficient p-value for each study.

Incorrect interpretations of the *p-value*

The *p-value* is not the probability that the null hypothesis (H_0) is true.

The *p-value* is not the probability of rejecting the null hypothesis when it is true (Type I error).

$(1 - p\text{-}value)$ is not the probability that the alternative hypothesis (H_1) is true.

The significance level of the test is not determined by the *p-value*; rather, it is arbitrarily determined prior to viewing the sample data after which the *p-value* is compared to it.

What does the *p-value* mean? It is calculated on the assumption that the null hypothesis is true; therefore, it is the probability that the null hypothesis is true.

How does one calculate the probability that the null hypothesis of the mean is true for a one-sided test using the *z*-Table?

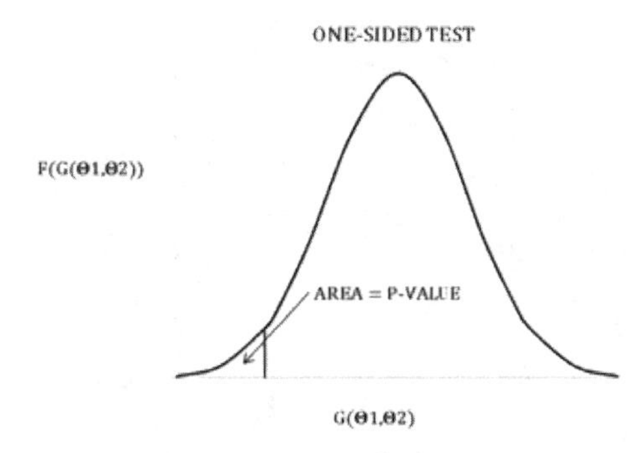

Example:
$H_0: H(\theta_1, \theta_2) = \mu - \mu_0 = 0,$
$H_1: H(\theta_1, \theta_2) = \mu - \mu_0 < 0$
Sample data is $n = 100, \bar{x} = 99.0,$
$\sigma = 10, \mu_0 = 100.$

The sample size is sufficiently large to use the *z*-Table.The probability that the null hypothesis is true? In other words, what is the *p-value*?

If the *z*-Table has an illustration with the tail on the left, then α will be the *p-value*.

$$G(\hat{\theta}_1, \theta_2) = \frac{\bar{x} - \mu_0}{\frac{\sigma}{\sqrt{n}}} = z_\alpha \sim N(0,1), \quad \frac{\bar{x} - \mu_0}{\frac{\sigma}{\sqrt{n}}} = \frac{99.0 - 100}{\frac{10.0}{\sqrt{100}}} = -1.00 = z_\alpha \rightarrow \alpha = 0.1587.$$

The probability that the null hypothesis is true is equal to the *p-value*.

$p\text{-}value = 0.1587$. Since $p\text{-}value > 0.10$, there is little or no evidence from the sample data to reject the null hypothesis.

How does one calculate the probability that the null hypothesis of the mean is true for a one-sided test using the *t*-Table?

Sample data is $n = 15, \bar{x} = 90.0, s = 12.9$. The sample size is not large enough to use the *z*-Table, so the *t*-Table must be used.

$H_0: H(\theta_1, \theta_2) = \mu - \mu_0 = 0,$

$H_1: H(\theta_1, \theta_2) = \mu - \mu_0 < 0$.The probability that the null hypothesis is true at the 10% significance level? In other words, what is the *p-value*? Note that the significance level is identified before the *p-value* is even calculated.

$$G(\hat{\theta}_1, \theta_2) = \frac{\bar{x} - \mu_0}{\frac{s}{\sqrt{n}}} = t_{1-\alpha}[n-1] \sim t[n-1]$$

$$\frac{90.0 - 100}{\frac{12.9}{\sqrt{15}}} = -3.00$$

If the t-Table contains no negative values for t, recall that because the t-Distribution is symmetrical about $t = 0$, $t_{1-\alpha}[n-1] = -t_\alpha[n-1]$. The table value of 3.00 for $(n-1) = 14$ corresponds to a tail area of 0.005.

$p\text{-}value = Pr(T \leq t_{1-\alpha}[n-1]) = Pr(T \leq -t_\alpha[14] = 3.00) = \alpha = 0.005.$

The probability that the null hypothesis is true is 0.005. Because $p\text{-}value < 0.10$, the null hypothesis is rejected at the 10% significance level.

How does one calculate the probability that the null hypothesis of a proportion is true for a one-sided test using the Z-Table?

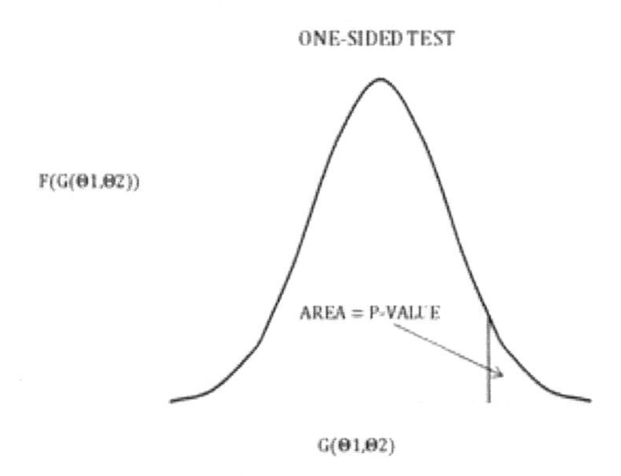

Example:
$H_0: H(\theta_1, \theta_2) = p - p_0 = 0,$

$H_1: H(\theta_1, \theta_2) = p - p_0 > 0$

Sample data is $n = 100, k = 54,$
$\sigma = 10, p_0 = 0.50, \hat{p} = k/n = 0.54,$

The sample size is sufficiently large to use the Z-Table, because $n\hat{p}, n(1-\hat{p}) \geq 10$.The probability that the null hypothesis is true? In other words, what is the *p-value*?

The value for z must be positive because the *p-value* is to the right of the mean, and the illustration in the table shows α is to the left of z, so the area to the right of z is $1-\alpha$.

$$G(\hat{\theta}_1, \theta_2) = \frac{\frac{k}{n} - p_0}{\sqrt{\frac{p_0(1-p_0)}{n}}} = z_{1-\alpha} \sim N(0,1),$$

$$\frac{\frac{k}{n} - p_0}{\sqrt{\frac{p_0(1-p_0)}{n}}} = \frac{0.54 - 0.50}{\sqrt{0.50(1-0.50)/100}} = 0.80 = z_{1-\alpha} \to 1-\alpha = 0.84, \alpha = 0.16\,.$$

The probability that the null hypothesis is true is equal to the *p-value*.

p-value $= 0.16$. Since *p-value* > 0.10, there is little or no evidence from the sample data to reject the null hypothesis.

How does one calculate the probability that the null hypothesis is true for a two-sided test?

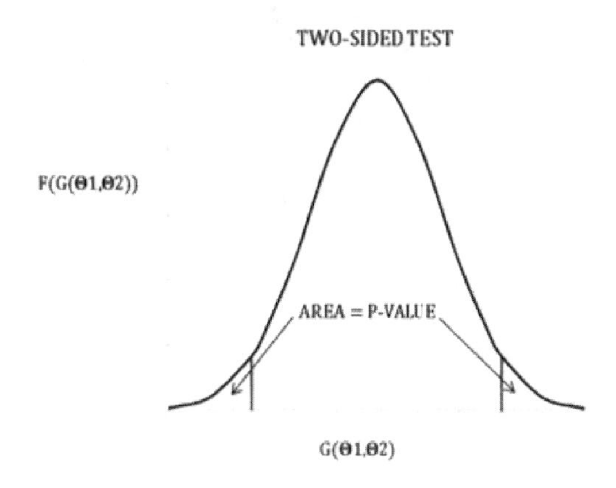

Suppose: To restate:

$H_0: p - p_0 = 0,$ $H_0: p - p_0 = 0,$

$H_1: p - p_0 \neq 0.$ $H_{1A}: p - p_0 > 0,$

$H_{1B}: p - p_0 < 0.$

The *p-value* will be the sum of the two tails in the illustration to the left, where the vertical lines are determined by calculating the transformed variables using the sample data.

The significance level is calculated before the *p-value* are calculated.

One would say that the null hypothesis is rejected at the X% significance level if *p-value* $<$ *significance level*, or the null hypothesis is not rejected at the X% significance level if *p-value* $>$ *significance level*.

Type I and Type II errors in a two-tailed hypothesis test

Suppose: $H_0: p - p_0 = 0$, $H_1: p - p_0 \neq 0$.

To restate: $H_0: p - p_0 = 0, H_{1A}: p - p_0 > 0, H_{1B}: p - p_0 < 0$.

The $significance\ level$ will be of the area under the curve labeled H_0 and to the left of the left vertical line plus the area under the same curve and to the right of the right vertical line. This sum will be the probability of a Type I Error. $Pr(Type\ I\ Error)$ is also known as the $a\text{-}Risk$.

The probability of a Type II Error will be the area under the curve labeled H_{1A} and to the left of the left vertical line plus the area under the curve labeled H_{1B} and to the right of the right vertical line. $Pr(Type\ II\ Error)$ is also known as the $\beta\text{-}Risk$. The probability of <u>not</u> committing a Type II Error is $(1 - \beta\text{-}Risk)$, also known as the power of the hypothesis test.

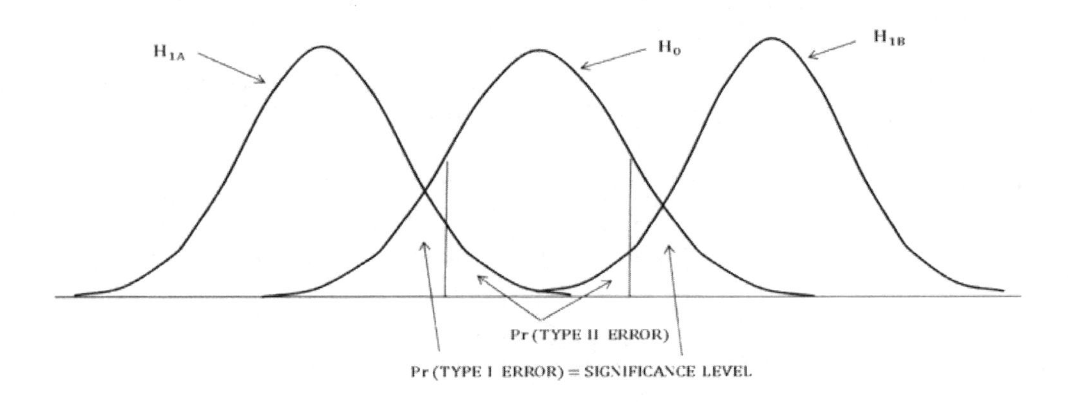

Type I Error

A Type I Error is the error of mistakenly rejecting a true null hypothesis. The probability of a Type I Error is $Pr(Type\ I\ Error) = \alpha\text{-}Risk = significance\ level$. It is the area under the curve labeled H_0 and to the left of the vertical line in the illustration below.

A Type I Error can be generally interpreted as an "error of excessive credulity" or a "False Positive"

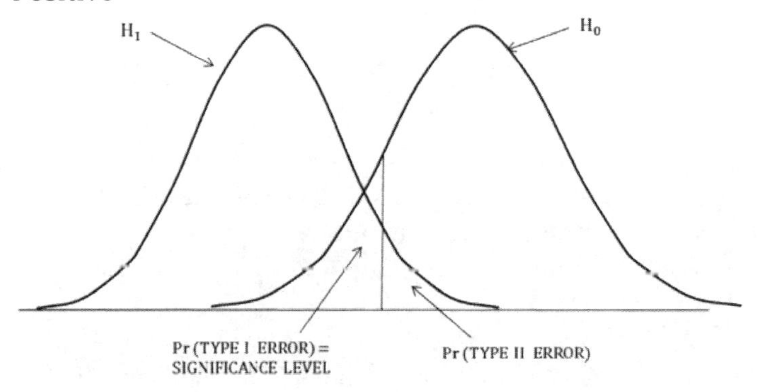

Significance level and statistical significance

The Significance Level is arbitrarily assigned a value by the statistician prior to the calculation of the p-value. It is expressed as a percentage.

Statistical Significance $=$ 100% $-$ Significance Level.

$$\alpha\text{-}Risk = \frac{(significance\ level)}{100\%}.$$

The sample data is transformed into G(observed) from which the p-value is calculated. The p-value and the α-Risk are each expressed as a fraction.

If p-value $< \alpha$-Risk, the null hypothesis is rejected. We say that the null hypothesis is rejected at the XX% significance level.
If p-value $> \alpha$-Risk, the null hypothesis is not rejected. We say that the null hypothesis is not rejected at the XX% significance level.

It is not required to specify a significance level. Note. Statistical significance is determined by comparing the resulting p-value to the a priori significance level, or alpha.

Significance Level

The *Significance Level* is arbitrarily created by the statistician prior to analysis of the data and the calculation of the *p-value*.

It establishes the arbitrary level for the *p-value*.

If *p-value* < *Significance Level*, the null hypothesis will be rejected.

If *p-value* > *Significance Level*, the null hypothesis will not be rejected.

Note. Significance Level is not the same as the *p-value*.

For example, suppose you find a p-value of .052, and you set your significance level at .05. Then, you will fail to reject the null hypothesis, claiming that there is not a difference between the two groups.

Type II Error

A Type II Error is the error of mistakenly failing to reject a false null hypothesis. The probability of a Type II Error is $Pr(Type\ II\ Error) = \beta\text{-}Risk$. It is the area under the curve labeled H_1 and to the right of the vertical line in the illustration below.

$(1 - \beta) = power\ of\ hypothesis\ test$

A Type II Error can be generally interpreted as an "error of excessive skepticism" or a "False Negative"

Power of Hypothesis Test

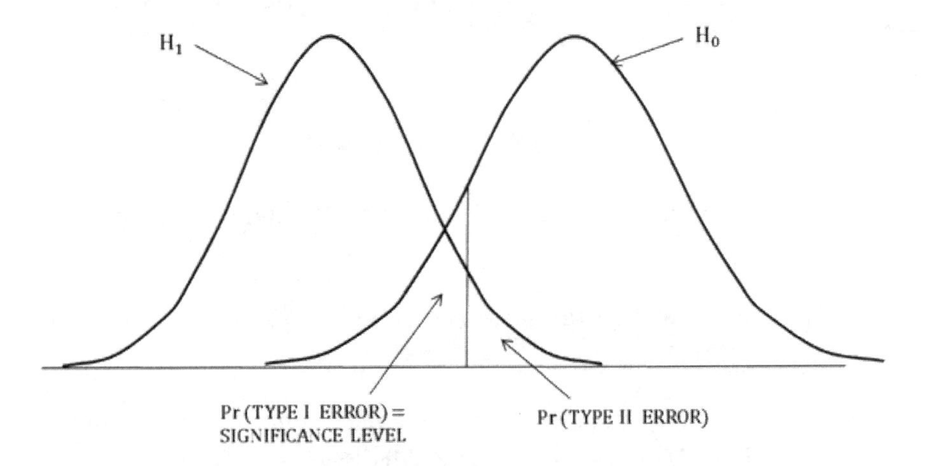

$Pr(Type\ II\ Error) = \beta\text{-}Risk$
$(1 - \beta) = power\ of\ hypothesis\ test$

The power of a hypothesis test is the probability that a Type II Error is <u>not</u> committed. The probability of a Type II Error, the $\beta\text{-}Risk$, is the area under the distribution curve representing the alternative hypothesis from the vertical line representing the boundary of the $\alpha\text{-}Risk$ to the closest tail.

Knowing the *Significance Level*, allows us to calculate the $\alpha\text{-}Risk$. But without knowing the average and variation of the distribution curve representing the alternative hypothesis, we cannot calculate the $\beta\text{-}Risk$ or the $power\ of\ hypothesis\ tes$.

Analyzing a two-sided hypothesis test for a proportion at the 10% significance level. What kind of null and alternate hypotheses would require a two-sided test?

One set of hypotheses that would require a two-sided test is a null hypothesis that says the distribution has a specific average, and an alternate hypothesis that says the distribution has an average that can be higher or lower. This would place an alternate distribution on either side of the null hypothesis distribution, thus generating a two-sided test.

Example: Suppose: Equivalently:

$H_0: p - p_0 = 0,\ \ H_0: p - p_0 = 0,$
$H_1: p - p_0 \neq 0.\ \ H_{1A}: p - p_0 > 0,$
$H_{1B}: p - p_0 < 0.$

Sample data is $n = 100, k = 54,$
$\sigma = 10, p_0 = 0.50, \hat{p} = k/n = 0.54,$

The sample size is sufficiently large to use the z-Table, because $n\hat{p}, n(1 - \hat{p}) \geq 10$.The probability that the null hypothesis is true at the 10% *significance level*?

The value for z must be both positive and negative, because this is a two-sided test. The illustration in the Z-Table shows α is to the left of z, so the area to the right of z is $1 - \alpha$.

$$G(\hat{\theta}_1, \theta_2) = \frac{\frac{k}{n} - p_0}{\sqrt{p_0(1-p_0)/n}} = \frac{0.54 - 0.50}{\sqrt{0.50(1-0.50)/100}} = 0.80 = z_{1-\alpha} \rightarrow 1-\alpha = 0.84, \alpha = 0.16\,.$$

The probability that the null hypothesis is true is equal to the *p-value.*

$p\text{-}value = 2(0.16) = 0.32$. Since $p\text{-}value > 0.10$, there is little or no evidence from the sample data to reject the null hypothesis.

How does one calculate the hypothesis test for a difference between two proportions using the *z*-Table

By the Central Limit Theorem, if k_1 and k_2 each have binomial distributions $b_1(k_1, n_1, p_1)$ and $b_2(k_2, n_2, p_2)$, and the $Unbiased\ Estimator\ of\ Mean = (\hat{p}_1 - \hat{p}_2)$, and the $Unbiased\ Estimator\ of\ Var(p_1 - p_2) = \hat{p}_1(1-\hat{p}_1)/n_1 + \hat{p}_2(1-\hat{p}_2)/n_2$, then

$$z = \frac{estimator\ of\ pop.variable - pop.variable}{estimate\ of\ variance\ of\ pop.variable} = \frac{(\hat{p}_1 - \hat{p}_2) - (p_1 - p_2)}{\sqrt{\hat{p}_1(1-\hat{p}_1)/n_1 + \hat{p}_2(1-\hat{p}_2)/n_2}}$$

converges in distribution as $n \rightarrow \infty$ to $N(0,1)$.

Example:

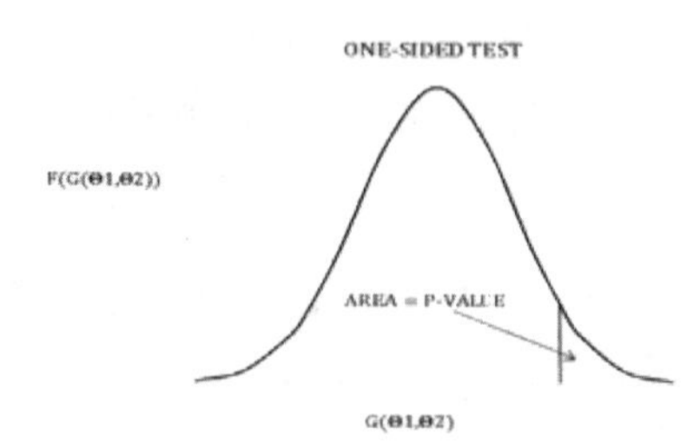

$H_0: H(\theta_1, \theta_2) = p_1 - p_2 = 0,$
$H_1: H(\theta_1, \theta_2) = p_1 - p_2 > 0.$

Sample data is
$n_1 = 85, k_1 = 50,\ \hat{p}_1 = k_1/n_1 = 0.59,$
$n_2 = 100, k_2 = 54, \hat{p}_2 = k_2/n_2 = 0.54,$

The sample size is sufficiently large to use the *z*-

The value for z must be positive because $(\hat{p}_1 - \hat{p}_2) - (p_1 - p_2) > 0$, and the illustration in the Z-table shows α is to the left of z, so the area to the right of z is $1 - \alpha$.

$$G(\hat{\theta}_1, \theta_2) = \frac{(\hat{p}_1 - \hat{p}_2) - (p_1 - p_2)}{\sqrt{\hat{p}_1(1-\hat{p}_1)/n_1 + \hat{p}_2(1-\hat{p}_2)/n_2}} = \frac{0.05 - 0}{\sqrt{0.59(1-0.59)/85 + 0.54(1-0.54)/100}} = 0.68 = z_{1-\alpha} \rightarrow 1-\alpha = 0.47, \alpha = 0.53\,.$$

The probability that the null hypothesis is true is equal to the *p-value.*

$p\text{-}value = 0.53$. Since $p\text{-}value > 0.10$, there is little or no evidence from the sample data to reject the null hypothesis.

One-tailed hypothesis for the difference between two unpaired means using the Z-Table. The example also demonstrates the improvement in the *power of the test* (probability of correctly accepting H_1) by increasing the sample size.

$H_0: H(\theta_1, \theta_2) = \mu_X - \mu_Y = 0, H_1: H(\theta_1, \theta_2) = \mu_X - \mu_Y = -2.00$

$\bar{X} = 20.00, \bar{Y} = 22.85, n_X = 20, n_Y = 25, s_X^2 = 16.00, s_Y^2 = 12.00$.The probability of making a Type I Error? What is the probability of making a Type II Error assuming $Var[X - Y] = Var[X] + Var[Y] = s_x^2 + s_y^2$?

$[n_X + n_Y - 2] = 43 > 30$, so the Z-Table may be used.

$$G(\hat{\theta}_1, \theta_2) = \frac{(\bar{x} - \bar{y}) - (\mu_x - \mu_y)}{\sqrt{s_x^2/n_x + s_y^2/n_y}} = z_\alpha \sim N(0,1),$$

$$= \frac{(20.00 - 22.85) - (0)}{\sqrt{16.00/20 + 12.00/25}} = -2.52 = z_\alpha \rightarrow \alpha = 0.0059.$$

This is a one-tailed test because $H_1: \mu_X - \mu_Y > 0$. Since a *significance level* has not been specified, $Pr(Type\ I\ Error) = \alpha = 0.0059 = p\text{-}value$, and H_0 is rejected since $p\text{-}value < 0.05$. Assuming H_1 is true,

$$\frac{(\bar{x} - \bar{y}) - (\mu_x - \mu_y)}{\sqrt{s_x^2/n_x + s_y^2/n_y}} = \frac{(20.00 - 22.85) - (-2.00)}{\sqrt{16.00/20 + 12.00/25}} = -0.75 = z_\beta \rightarrow \beta = 0.2266$$

Therefore, $Pr(Type\ I\ Error) = \beta = 0.23$ and the *power of the test* (probability of correctly accepting H_1) is $(1 - \beta) = 0.73$. Not that good! If the sample size is increased to $n_X = n_Y = 80$,

$Pr(Type\ I\ Error) = \beta = 0.075$ and the *power of the test* (probability of correctly accepting H_1) is $(1 - \beta) = 0.925$. A tremendous improvement by increasing the sample size!

Two-sided hypothesis test for the difference between two unpaired means using the T-Table

$H_0: H(\theta_1, \theta_2) = \mu_X - \mu_Y = 0, H_1: H(\theta_1, \theta_2) = \mu_X - \mu_Y \neq 0$
$\bar{X} = 20.00, \bar{Y} = 22.85, n_X = 12, n_Y = 10, s_X^2 = 4.00, s_Y^2 = 5.00.$

The probability of making a Type I Error?

$[n_X + n_Y - 2] = 20 < 30$, so the t-Table must be used.

$$G(\hat{\theta}_1, \theta_2) = \frac{(\bar{X} - \bar{Y}) - (\mu_X - \mu_Y)}{\sqrt{\left(\frac{(n_X - 1)s_X^2 + (n_Y - 1)s_Y^2}{n_X + n_Y - 2}\right)}\sqrt{\left(\frac{1}{n_X} + \frac{1}{n_Y}\right)}} = t_\alpha[n_X + n_Y - 2] \sim t[n_X + n_Y - 2],$$

$$= \frac{(20.00 - 22.85) - (0)}{\sqrt{\left(\frac{(12-1)4.0 + (10-1)5.0}{12+10-2}\right)}\sqrt{\left(\frac{1}{12}+\frac{1}{10}\right)}} = -3.155 = t_{\alpha}[20].$$

Note: If negative values are not in the T-Table and the illustration has the tail on the right, remove the negative sign, replace α with $1 - \alpha$ and proceed:
$3.155 = t_{1-\alpha}[20] \rightarrow (1 - \alpha) = 0.9975$ and $\alpha = 0.0025$.

The table value of 3.155 for $(n_X + n_Y - 2) = 20$ corresponds to a tail area of 0.0025.

This is a two-tailed problem, because $H_1: \mu_X - \mu_Y \neq 0$ means $\mu_X - \mu_Y > 0$ or $\mu_X - \mu_Y < 0$.

Therefore, the $p\text{-}value = Pr(T > t_{\alpha}[20], T < -t_{\alpha}[20]) = 2\alpha = 0.005$.

The probability that H_0 is true is 0.005. Because the $p\text{-}value$ $0.005 < 0.050$, H_0 is rejected. Because a *significance level* was not specified, the $Pr(Type\ I\ Error) = 0.005$. We cannot calculate $Pr(Type\ II\ Error)$ because H_1 does not identify a specific value for the true average of the difference between the two populations $(X - Y)$: $E[X - Y] = E[X] + E[Y] = \mu_X - \mu_Y$, or the true variation of $(X - Y)$: $Var[X - Y] = Var[X] + Var[Y] = \sigma_x^2 + \sigma_y^2$.

One-sided hypothesis test for the difference between two paired means using the T-Table

$H_0: \mu_D = 0,\ H_1: \mu_D > 0.$
$\bar{D} = 2.18, n = 12, s_{\bar{D}}^2 = 4.00$.The probability of making a Type I Error?

$$\frac{\bar{D} - \mu_D}{s_{\bar{D}}} = t_{\alpha}[n-1] \sim t[n-1], \text{ where } X \sim N(\mu_X, \sigma_X^2),\ Y \sim N(\mu_Y, \sigma_Y^2),$$

$$\mu_D = E(X_i - Y_i),\ s_{\bar{D}}^2 = \frac{1}{n-1}\sum_{i=1}^{i=n}(x_i - y_i)^2.$$

$$\frac{\bar{D} - \mu_D}{s_{\bar{D}}} = \frac{2.18 - 0}{2.00} = 1.09 = t_{\alpha}[11] \rightarrow \alpha = 0.15$$

Therefore, $p\text{-}value = Pr(T > t_{\alpha}[11]) = \alpha = 0.15$.

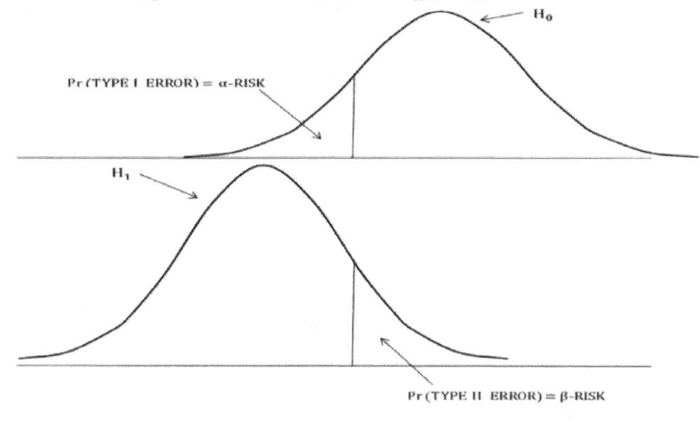

The probability that H_0 is true is 0.15. Because the $p\text{-}value = 0.15 > 0.10$, H_0 is not rejected on the evidence of the sample data.

Power of hypothesis test in terms of the probability of a Type II Error and an illustration

A Type II Error is the error of mistakenly failing to reject a false null hypothesis.

$Pr(Type\ II\ Error) = \beta\text{-}Risk, (1-\beta) = power\ of\ hypothesis\ test.$

The curve representing the null hypothesis H_0 is false. Therefore, the calculation of the transformed variable from the sample data (the vertical line in the illustration) is really coming from the curve representing the alternate hypothesis H_1. If the area for the probability of accepting H_0 and rejecting H_1 is the area β under the distribution labeled H_1, then the probability of accepting the true hypothesis H_1 is $(1-\beta)$. In other words, the *power of the hypothesis test* is the probability of accepting the correct hypothesis.

Pearson's Chi-Square Test, and why does it not have the general form of a Chi-Square Distribution?

Recall the general form of a variable that has a Chi-Square Distribution,

$\sum_{i=1}^{i=v} Z_i^2 \sim \chi^2(v)$, where $Z_i \sim N(0,1)$ for all i.

Pearson's Chi-Square Test, however, is not of this form. It is an approximation that tests the hypothesis that the frequency distribution of events in a sample have a certain distribution. It is particularly suited for studying the goodness of fit to an assumed distribution and independence of categorical events in a contingency table.

General Form: The general form of Pearson's Chi-Square Test is

$$X^2 = \sum_{i=1}^{n_{CELLS}} \frac{(O_i - E_i)^2}{E_i}, \lim_{n\to\infty} \sum_{i=1}^{n} \frac{(O_i - E_i)^2}{E_i}$$
$$= \chi_\alpha^2[n_{CELLS} - n_{PARA} - 1] \sim \chi^2[n_{CELLS} - n_{PARA} - 1],$$

n_{CELLS} is the number of cells or categories in the table,

$(n_{PARA} + 1)$ is the number of degrees of freedom lost by calculating the n_{PARA} parameters for the assumed distribution (for example, $n_{PARA} = 2$ for mean and variation of a normal distribution),

E_i is the expected value for event A_i,
O_i is the observed value for event A_i.
Pearson's test statistic, X^2, asymptotically approaches a $\chi_\alpha^2[n_{CELLS} - n_{PARA} - 1]$ distribution.

Requirements and assumptions of Pearson's Chi-Square Test

The requirements and assumptions for Pearson's Chi-Square Test,

$$X^2 = \sum_{i=1}^{n_{CELLS}} \frac{(O_i - E_i)^2}{E_i},$$

are as follows,

Requirements: If the events are A_i, $i = 1,2 \dots n_{CELLS}$, the events must be mutually exclusive and their total probability must be unity,

$$A_i \cap A_j \text{ for } i \neq j, \text{ and } \sum_{i=1}^{i=n_{CELLS}} P(A_i) = 1,$$

where $P(A_i)$ is the probability of event A_i.

Assumptions:

Random sampling.

The total sample size (sum of all elements in each set A_i) must be "sufficiently large".

Sample size of each cell (number of elements of each set A_i) must be ≥ 5.

The observations are independent.

Pearson's Chi-Square Test for goodness of fit

Objective: Determine if an observed pattern of data fits a particular distribution. A small *p-value* will reject H_0 that the categorical variable has the selected distribution.

There is one cell for each sample category $degrees\ of\ freedom = (\#\ cells - 1)$, because if we know the total of all observations and the observations in all but one of the cells, we can calculate the observations in the remaining cell

Example: Assume a discrete uniform distribution. There are 60 tosses of a die. The observed (O_i) and expected (E_i) values are below. Is the die fairly balanced?
$O_i = (1/6)(60) = 10$, where O_i represents the number of times the face with i dots appears.

E_1	E_2	E_3	E_4	E_5	E_6	→	10	10	10	10	10	10
O_1	O_2	O_3	O_4	O_5	O_6	→	8	1	7	6	1	1

The value in each cell is ≥ 5, the observations are independent and the sampling is random.

$$X^2 = \sum_{i=1}^{n_{CELLS}} \frac{(O_i - E_i)^2}{E_i} = \frac{2^2}{10} + \frac{4^2}{10} + \frac{3^2}{10} + \frac{4^2}{10} + \frac{3^2}{10} + \frac{1^2}{10} = 5.5, df = 5$$

$\chi_{\alpha}^2[5] = 5.5 \rightarrow \alpha = 0.36$, Since *p-value* =0.36>0.1, there is no evidence to reject the null hypothesis that the die is fairly balanced.

Pearson's Chi-Square Test for independence

Objective: Select a sample from a single population that possesses two categorical variables and determine if there is an association between the variables. A small *p-value* will reject the H_0 that there is no association between the variables. There is a row for each level of the sample category, and there is a column for each level of the categorical variable. The total number of cells is the number of rows times the number of columns (rc). Loss of degrees of freedom is $(r + c - 1)$ because if we know the total of the row (column) sums and all but one of the column (row) sums, we can compute the remaining column (row) sum.

$$degrees\ of\ freedom = rc - (r + c - 1) = (r - 1)(c - 1),$$

$$A_{ij} = A_i \cap A_j, P(A_i|A_j) = \frac{P(A_i \cap A_j)}{P(A_j)},$$

If A_i and A_j are independent, $P(A_i|A_j) = P(A_i)$, $P(A_i \cap A_j) = P(A_i)P(A_j)$, and in each cell the expected value is, $A_{ij} = NP(A_i)P(A_j)$, where N is the total number of observations.

If E_i is the expected value for event A_i, and O_i is the observed value for event A_i, then $E_{ij} = NP(O_i)P(O_j)$

A contingency table of expected events, E_{ij}, and a contingency table of observed events, O_{ij}, are constructed. Next, the following is calculated,

$$X^2 = \sum_{j=1}^{j=c}\sum_{i=1}^{i=r}\frac{(O_{ij} - E_{ij})^2}{E_{ij}}.$$

This is an estimate of $\chi^2_\alpha[(r-1)(c-1)]$. H_0 is rejected or not based on *p-value* = α.

Pearson's Chi-Square Test for independence of variables in the same population

Example: 200 people are asked to rank three candidates (A, B and C) in order of preference.

$$P(E_A) = \frac{67}{200}, P(E_B) = \frac{65}{200}, P(E_C) = \frac{68}{200}, P(E_1) = \frac{50}{200}, P(E_2) = \frac{95}{200}, P(E_3) = \frac{55}{200},$$

H_0: No interaction between candidates and preference. $E_{ij} = NP(O_i)P(O_j)$:
$E_{A1} = NP(E_A)P(E_1) = 16.8$, $E_{A2} = NP(E_A)P(E_2) = 31.8$, $E_{A3} = NP(E_A)P(E_3) = 18.4$,
$E_{B1} = NP(E_B)P(E_1) = 16.3$, $E_{B2} = NP(E_B)P(E_2) = 30.9$, $E_{B3} = NP(E_B)P(E_3) = 17.9$,
$E_{C1} = NP(E_C)P(E_1) = 17.0$, $E_{C2} = NP(E_C)P(E_2) = 32.3$, $E_{C3} = NP(E_C)P(E_3) = 18.7$,

	1st	2n	3rd
A	E_{11}	E_{12}	E_{13}
B	E_{21}	E_{22}	E_{23}
C	E_{31}	E_{32}	E_{33}

→

	1st	2n	3rd
A	16.	31.	18.
B	16.	30.	17.
C	17.	32.	18.

	1st	2n	3rd
A	O_{11}	O_{12}	O_{13}
B	O_{21}	O_{22}	O_{23}
C	O_{31}	O_{32}	O_{33}

→

	1st	2n	3rd	
A	35	20	12	67
B	10	45	10	65
C	5	30	33	68
	50	95	55	20

$\chi^2_\alpha[(r-1)(c-1)] = \chi^2_\alpha[4] = 29.5 \rightarrow \alpha < 0.001 =$ *p-value* . H_0 is rejected.

Pearson's Chi-Square Test for homogeneity of proportions

Objective: Select a sample from each of two or more populations and compare the distribution of a single categorical variable among the different populations. A small *p-value* will reject the H_0 that the populations have the same distribution for the different levels of the categorical variable.

Additional assumption is that the original population be much larger than the sample size.

There is a row for each sample category. Each sample is assumed to come from a different population. There is a column for the number in each proportion category.

Loss of degrees of freedom is $(r + c - 1)$ because if we know the total of the row (column) sums and all but one of the column (row) sums, we can compute the remaining column (row) sum.

$$degrees\ of\ freedom = rc - (r + c - 1) = (r - 1)(c - 1).$$

Pearson's Chi-Square Test to determine if a variable has the same distribution in different populations

Example: 200 people from three countries (A, B, and C) are asked to rank their opinion of capitalism (1=good, 2=fair, 3=poor).

$$P(E_A) = \frac{67}{200}, P(E_B) = \frac{65}{200}, P(E_C) = \frac{68}{200}, P(E_1) = \frac{50}{200}, P(E_2) = \frac{95}{200}, P(E_3) = \frac{55}{200},$$

H_0: Categorical variable, opinion of capitalism, has the same distribution in all three countries.

$$E_{A1} = NP(E_A)P(E_1) = 16.8,\ E_{A2} = NP(E_A)P(E_2) = 31.8,\ E_{A3} = NP(E_A)P(E_3) = 18.4,$$
$$E_{B1} = NP(E_B)P(E_1) = 16.3,\ E_{B2} = NP(E_B)P(E_2) = 30.9,\ E_{B3} = NP(E_B)P(E_3) = 17.9,$$
$$E_{C1} = NP(E_C)P(E_1) = 17.0,\ E_{C2} = NP(E_C)P(E_2) = 32.3,\ E_{C3} = NP(E_C)P(E_3) = 18.7,$$

	1	2	3
A	E_{11}	E_{12}	E_{13}
B	E_{21}	E_{22}	E_{23}
C	E_{31}	E_{32}	E_{33}

→

	1	2	3
A	16.	31.	18.
B	16.	30.	17.
C	17.	32.	18.

	1	2	3
A	O_{11}	O_{12}	O_{13}
B	O_{21}	O_{22}	O_{23}
C	O_{31}	O_{32}	O_{33}

→

	1	2	3	
A	35	20	12	67
B	10	45	10	65
C	5	30	33	68
	50	95	55	20

$\chi_\alpha^2[(r - 1)(c - 1)] = \chi_\alpha^2[4] = 29.5 \rightarrow \alpha < 0.001 = p\text{-}value$. H_0 is rejected.

Outline for deriving the confidence interval for the slope of a least-squares regression line

$$Slope = \frac{Cov(X,Y)}{Var(X)},$$

$Var(X) = E[\{X - E(X)\}^2]$, $Cov(x,y) = E[\{X - E(X)\}\{Y - E(Y)\}]$.

The unbiased estimators of $Var(X)$ and $Cov(x,y)$ are

$$\widehat{Var}(X) = \frac{1}{n-1}\sum_{i=1}^{i=n}(x_i - \bar{x}_n)^2, \widehat{Cov}(x,y) = \frac{1}{n-1}\sum_{i=1}^{i=n}(x_i - \bar{x}_n)(y_i - \bar{y}_n).$$

$$\widehat{Slope} = \frac{\sum_{i=1}^{i=n}(x_i - \bar{x}_n)(y_i - \bar{y}_n)}{\sum_{i=1}^{i=n}(x_i - \bar{x}_n)^2}.$$

$Var(Slope) = E[\{Slope - E(Slope)\}^2]$,

$$\widehat{Var}(Slope) = \frac{\widehat{Var}(\hat{y})}{(n-1)\widehat{Var}(x)} = \frac{\frac{1}{n-2}\sum(y_i - \hat{y}_i)^2}{\sum(x_i - \bar{x}_n)^2}.$$

$$Slope \pm t_{\alpha/2}[n-2]\sqrt{\frac{\frac{1}{n-2}\sum(y_i - \hat{y}_i)^2}{\sum(x_i - \bar{x}_n)^2}}$$ at the $(1-\alpha)100\%$ Confidence Level.

Hypothesis test on the slope of a least squares line. The distribution used in the test is the T-Distribution.

The set of data $\{(x_i, y_i)\}$ is in the left table. The terms and their values using the functions in EXCEL are in the right table.

xi	yi
10	8
11	10
12	8
13	16
14	13
15	19
16	15
17	30
18	20
19	28
20	24

$$\frac{\widehat{Slope} - Slope}{\frac{\widehat{Var}(Slope)}{\sqrt{(n-1)}}} = t_\alpha[n-2] \sim t[n-2],\ \widehat{Slope} = 1.981818,$$

$$\widehat{Var}(Slope) = \frac{\widehat{Var}(\hat{y})}{(n-1)\widehat{Var}(x)} = \frac{SE(\hat{y})^2}{(n-1)SE(x)^2} = \frac{4.089405^2}{(10)3.316625^2} = 0.152$$

If H_0: $Slope = 0$,

$$\frac{\widehat{Slope} - Slope}{\frac{\widehat{Var}(Slope)}{\sqrt{(n-1)}}} = \frac{1.981818 - 0}{0.152/\sqrt{10}} = 4.12 = t_\alpha[9] \rightarrow \alpha = p\text{-}value = 0.0013.$$ H_0 is rejected.

Correlation

A correlation is a relationship between bivariate data typically observed in a scatter plot.

Example: Bivariate data is recorded for the volume of ice cream sales and the number of drowning deaths. A strong positive correlation is found, even though no there is no reasonable causation. However, a well-designed experimental study would look for other variables for a causal relationship and consequently reveals causation in the correlation between the outdoor temperature and the volume of ice cream sales, as well as between the outdoor temperature and the number of drowning deaths.

A correlation can be strong or weak, positive- or negative linear, non-linear or none.

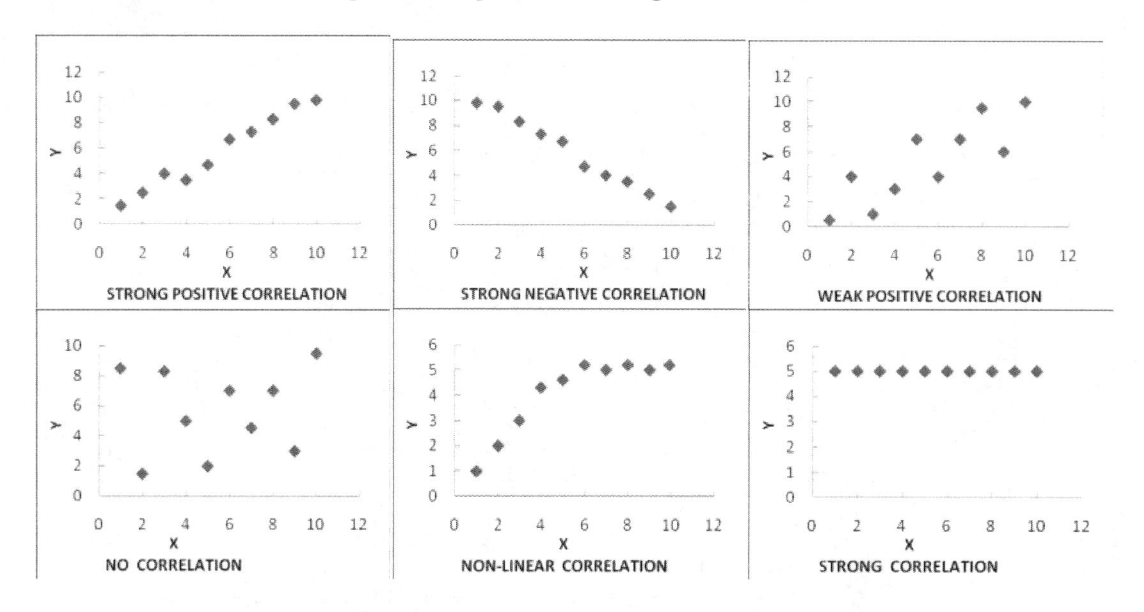

Note. Any correlation found in such data does not imply causation. Instead, the experimenter can claim that there is a relationship between the variables, with the degree of the relationship supported by the value of *r*, the correlation coefficient.

Correlation does not imply causation.

Example: Bivariate data is recorded for the volume of ice cream sales and the number of drowning deaths.

X_3 —— $S(X_3)$ ⟶ X_4
X_3 ⟵ $S^{-1}(X_4)$ —— X_4
X_3 —— $P(X_3)$ ↓ X_2
X_2 —— $D(X_2)$ ⟶ X_1

where X_1 is the number of drowning deaths, X_2 is the number of people in the water, X_3 is the month of the year, X_4 is the amount of ice cream sales, $D(X_2)$ is the number of drowning deaths as a function of the number of people in the water, $P(X_3)$ is the number of people in the water as a function of the month of the year, $S(X_3)$ is ice cream sales as a function of the month of the year, and $S^{-1}(X_4)$ is the month of the year as a function of ice cream sales.

$$X_1 = D(X_2) = D\big(P(X_3)\big) = D\big(P(S^{-1}(X_4))\big)$$

The above equation indicates that the number of drowning deaths is a function of ice cream sales. However, despite this equation, X_4 is not an independent variable in the experimental sense because ice cream sales is not a controlled variable. If ice cream sales were controlled while keeping the season constant, no amount of ice cream sold in winter could result in any drowning deaths. Correlation does not imply causation; it simply provides the strength of the relationship between two or more variables, revealing how close they are related or correlated. Other statistical tests, such as t-tests and ANOVAs are used to test the effect of an independent variable on a dependent variable.

Least-squares regression line

Given a scatter plot of a sample of bivariate continuous data of the form (x, y), we arbitrarily choose y to be the dependent variable, x to be the independent variable, and we try to simulate the data with a linear equation, $y = \alpha + \beta x$, that will pass a straight line through the data. Each experimental data point (x_i, y_i) is now predicted to be $(x_i, \alpha + \beta x_i)$. On the scatter plot, the vertical distance between each experimental data point and its predicted data point is $|y_i - (\alpha + \beta x_i)|$. We want to fit the line to minimize these distances. One approach is the Method of Least Squares which adjusts the α and β so that the sum of the square of each distance is minimized. The straight line is now called the "least-squares regression line".

Correlation coefficient and the coefficient of determination

The linear equation for the least-squares regression line of continuous bivariate (x_i, y_i) data minimizes the sum of the squares of the distances between the experimental value for each y_i and its predicted value $\hat{y}_i$. The equation for the predicted values is $\hat{y}_i = \bar{y} + r\frac{s_y}{s_x}(x_i - \bar{x})$, where $\bar{y}$ and $\bar{x}$ are the average values for x and y, s_y and s_x are the standard deviations (square root of the variances) for x and y, and r is the correlation coefficient.

$$r = \frac{1}{n-1}\frac{\sum_{i=1}^{i=n}(x_i - \bar{x})(y_i - \bar{y})}{s_x s_y}.$$

The correlation coefficient, r, is between -1 and 1.

The coefficient of determination is r^2, so its value is between 0 and 1. Another way to write the coefficient of determination is, $r^2 = \frac{\sum_{i=1}^{i=n}(\hat{y}_i - \bar{y})^2}{\sum_{i=1}^{i=n}(y_i - \bar{y})^2} = \frac{\text{explained variance}}{\text{total variance}}$

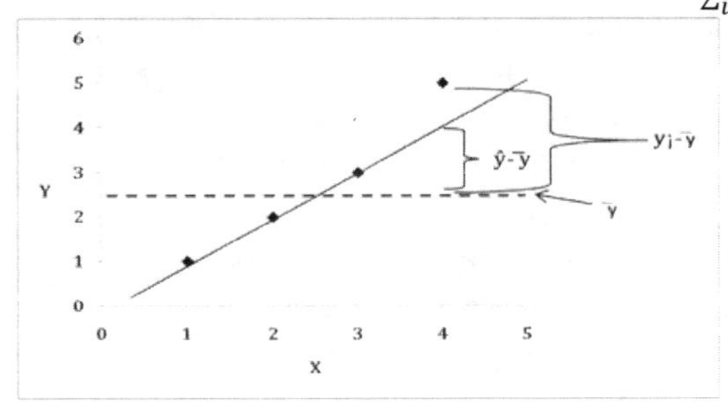

Leverage and influential points

A leverage point is a data point that is extremely high or extremely low in the data set and which has few, if any, other data points nearby, for which reason the fitted model will predict exceptionally well.

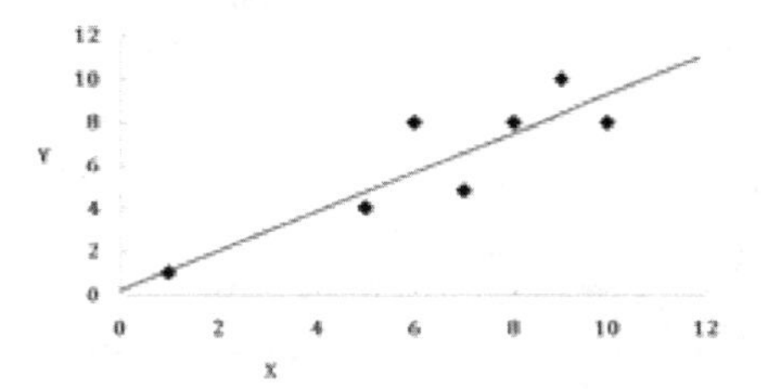

The data point (1,1) is highly leveraged, because it is an extreme value with no nearby data points, and, as a result, is exceptionally well predicted by the linear equation derived from the least-squares regression analysis.

An influential point is one whose presence causes a big change in the regression analysis calculation of the slope, α, and intercept, β, for the linear model, $\hat{y} = \alpha x + \beta$.

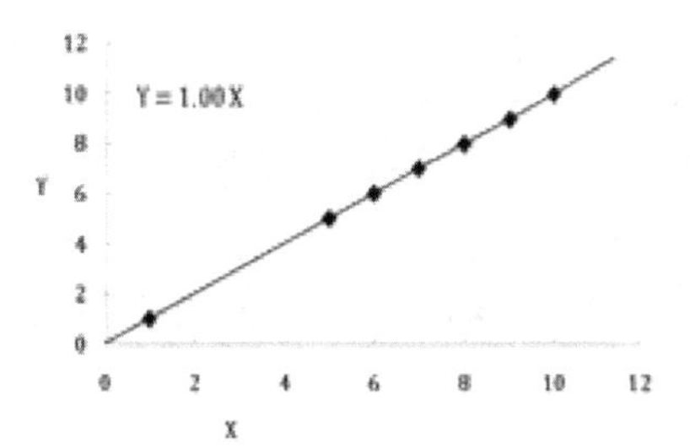

In the above two scatter plots, the data point (1,1) is a highly leveraged point, but it is not an influential point, because its presence does not change the slope or intercept. However, the data point (6,1) would obviously be an influential point. Influential points and highly leveraged points may or may not be outliers, depending on the definition of outlier. One definition is any value less than $Q_1 - 1.5(IQR)$ or any value greater than $Q_3 + 1.5(IQR)$.

Residual

For univariate data, the residual is the difference between a data point and the estimated mean, $x_i - \bar{x}$. For bivariate data that has undergone regression analysis, the residual is the difference between an experimental value, y_i, and the predicted value, $\hat{y_i}$, where $\hat{y_i} = \alpha x_i + \beta$. In a typical scatter plot with regression analysis, all data points (x_i, y_i) are plotted together with the straight line, $\hat{y_i} = \alpha x_i + \beta$. The assumptions of regression analysis regarding the residuals are that they: (i) have constant variance with respect to x, (ii) are uncorrelated with x, (iii) have zero mean over all x, and (iv) are normally distributed. In the residual plot below, the mean of the residual is zero, but the variance of the residual is dependent on x.

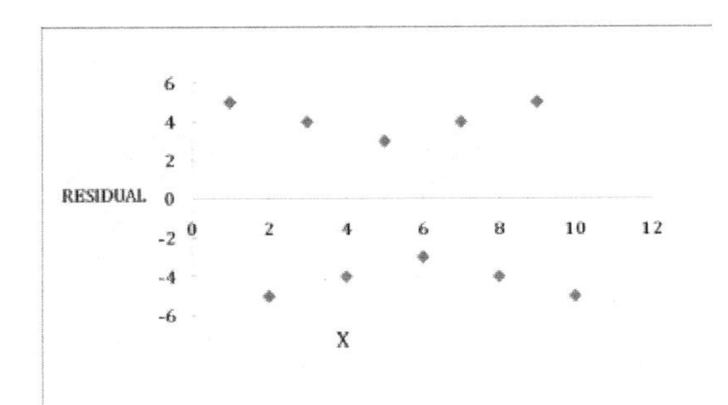

In the residual plot below, the mean of the residual is zero, the variance of the residual is independent on x, but the residual itself is dependent on x.

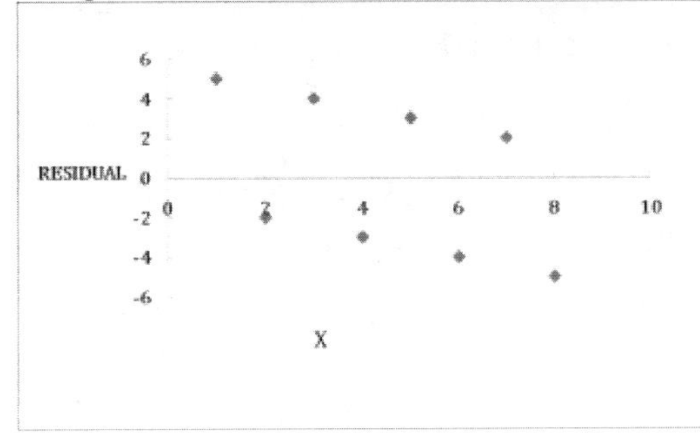

One-way and two-way table

An example of a one-way table of frequency distributions of univariate categorical variables with an ordinal scale of measure: One hundred reviewers are asked to rank a specific movie on an ordinal scale of measure from 1 to 5, where 1 is awful and 5 is outstanding. The one-way table includes rank, description and # of movie reviewers who assigned the movie a specified rank. Another example with nominal scale of measure is the number of cars of various color in a parking lot. The one-way table includes the color, the number of cars with that color and the cumulative frequency.

Category (with ordinal scale of measure)	Description	Frequency (# of Movie Reviewers)
1	awful	5
2	fair	20
3	good	25
4	excellent	30
5	outstanding	2

Category (with nominal scale of measure)	Frequency (# of cars)	Cumulative Frequency
red	10	10
blue	60	70
green	100	170
black	150	320
white	5	325

An example of a two-way table of frequency distributions of bivariate categorical variables is the number of cars and trucks of a specific color in a parking lot. Both categorical variables use a nominal scale of measure.

		Category #1 (Color)		
		red	black	green
Category #2 (Vehicle)	car	2	25	3
	truck	15	3	7

Tables used to record the relation between two or more categorical variables are called contingency tables.

Marginal and joint probability

The marginal probability, $P(X)$, is the probability of belonging in a specific subcategory. For example, in the contingency table below, let X be the categorical variable "Vehicle" and Y be the categorical variable "Color". The probability that a vehicle is a truck (subcategory of "Vehicle") is $P(X = truck)$ and is calculated,

$$P(truck) = \sum_Y P(truck, Y) = P(truck, yellow) + P(truck, grey) + P(truck, black)$$

$P(truck) = \frac{2}{79} + \frac{5}{79} + \frac{20}{79} = \frac{27}{79}$, where 79 is the total number of all vehicles of all colors.

The joint probability, $P(X, Y)$, is the probability of belonging to two specific subcategories. For example, the joint probability that a vehicle is a yellow truck, $P(X = truck, Y = yellow) = \frac{2}{79}$.

		Category #1 (Color)			TOTALS
		yellow	grey	black	
Category #2 (Vehicle)	truck	2	5	20	27
	car	2	25	3	30
	motorcycle	15	4	3	22
TOTALS		19	34	26	79

Marginal probability is also known as marginal relative frequency.

Joint probability is also known as joint relative frequency.

Conditional probability

The conditional probability, $P(X|Y)$, is the probability that X is a certain value on the condition that Y is a certain value.
(joint probability) = (conditional probability) x (marginal probability)

This is written as,
$P(X, Y) = P(X|Y)P(Y)$.

For example, in the contingency table below, let X be the categorical variable "Vehicle" and Y be the categorical variable "Color". The conditional probability, $P(truck|yellow)$ is calculated,
$P(truck, yellow) = P(truck|yellow)P(yellow)$.

In other words, (the probability that a vehicle in the parking lot is a yellow truck) = (the probability of the vehicle being a truck on the condition that the vehicle is yellow) X (the probability that the vehicle is yellow).

$P(truck, yellow) = \frac{2}{79}$, $P(yellow) = \frac{19}{79}$, therefore,
$P(truck|yellow) = \frac{P(truck, yellow)}{P(yellow)} = \frac{2}{19}$.

		Category #1 (Color)			TOTALS
		yellow	grey	black	
Category #2 (Vehicle)	truck	2	5	20	27
	car	2	25	3	30
	motorcycle	15	4	3	22
TOTALS		19	34	26	79

Conditional probability is also known as conditional relative frequency.

Transformation of data

Transformation of data is accomplished by converting each data point x_i into another data point x_i' by inserting it into an equation, $x_i' = f(x_i)$. Transformation of a random variable is also a random variable. For example, if numbers are randomly created by a random number generator, and then a constant is added to each number, they are still random numbers. Likewise, if each number is converted to its log, those are also random numbers.

Transformation of data is indicated when
(i) the population from which the sample is drawn is not normally distributed and a transformation would result in normalization, (ii) a scatter plot has most of the data located in one region and a transformation would result in spreading the data out, making it easier to visualize,

(iii) transformation would simplify interpretation (i.e., non-linear correlation in scatter plot is transformed into a linear correlation, transforming with the inverse function to convert miles/gallon to gallons /mile), (iv) a regression analysis shows the residual $(\hat{y}_i - y_i)$ changes with x_i in the equation $\hat{y}_i = \alpha x_i + \beta$.

Types of transformations include linear, logarithmic, square root, and logit, to name a few.

One of the assumptions of regression analysis is that the residuals cannot be a function of the variable. Possible transformation functions for x_i include the reciprocal, the reciprocal square root, the log and the square root. The transformation from x_i to x_i' makes the residuals random with respect to the transformed variable.

Linear regression analysis assumes a linear model $(\hat{y}_i = \alpha x_i + \beta)$ and calculates the residual $(\hat{y}_i - y_i)$ for each x_i. In the scatter plot below left, the residual decreases with increasing x_i.

In the scatter plot below right, the transformation of x_i with the appropriate function has stabilized the residual with respect to the variable.

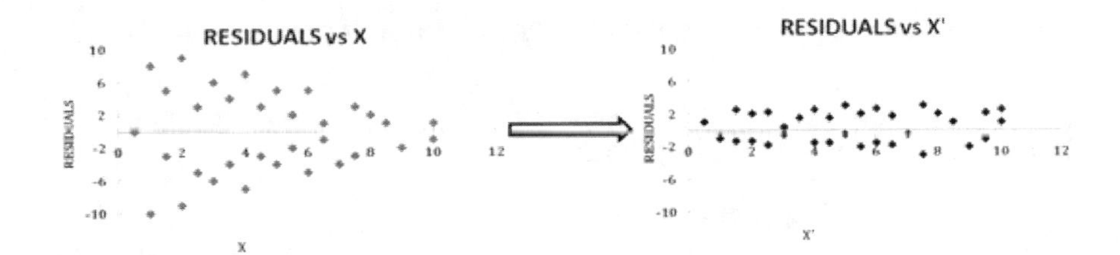

The scatter plot below left has been linearized by the transformation $x \to lnx$.

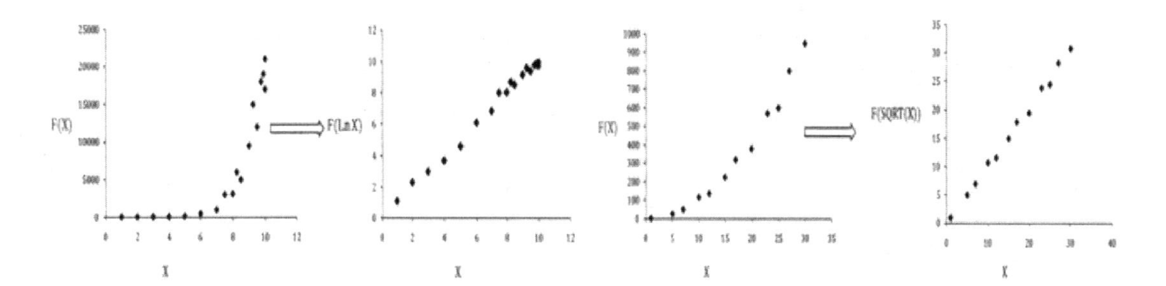

The scatter plot above right has been linearized by the transformation $x \to \sqrt{x}$.
The scatter plot below has been linearized by the transformation $x \to 1/x$.

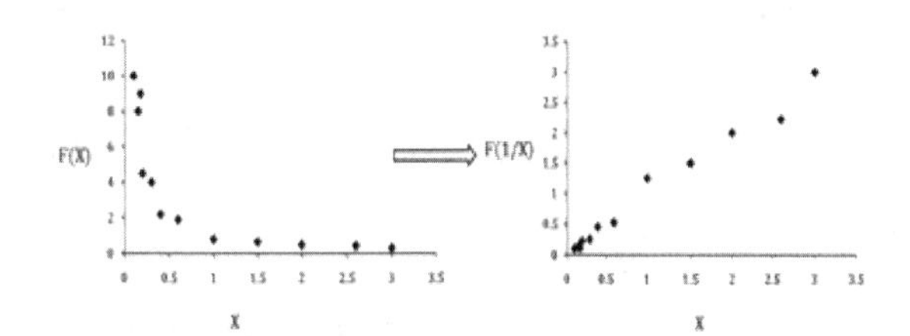

Indicate the general outline for deriving the confidence interval for the slope of a least-squares regression line

$$Slope = \frac{Cov(X,Y)}{Var(X)},$$

$Var(X) = E[\{X - E(X)\}^2]$, $Cov(x,y) = E[\{X - E(X)\}\{Y - E(Y)\}]$.

The unbiased estimators of $Var(X)$ and $Cov(x,y)$ are

$$\widehat{Var}(X) = \frac{1}{n-1}\sum_{i=1}^{i=n}(x_i - \bar{x}_n)^2, \widehat{Cov}(x,y) = \frac{1}{n-1}\sum_{i=1}^{i=n}(x_i - \bar{x}_n)(y_i - \bar{y}_n).$$

$$\widehat{Slope} = \frac{\sum_{i=1}^{i=n}(x_i - \bar{x}_n)(y_i - \bar{y}_n)}{\sum_{i=1}^{i=n}(x_i - \bar{x}_n)^2}.$$

$Var(Slope) = E[\{Slope - E(Slope)\}^2]$,

$$\widehat{Var}(Slope) = \frac{\widehat{Var}(\hat{y})}{(n-1)\widehat{Var}(x)} = \frac{\frac{1}{n-2}\Sigma(y_i - \hat{y}_i)^2}{\Sigma(x_i - \bar{x}_n)^2}.$$

$$Slope \pm t_{\alpha/2}[n-2]\sqrt{\frac{\frac{1}{n-2}\Sigma(y_i - \hat{y}_i)^2}{\Sigma(x_i - \bar{x}_n)^2}} \text{ at the } (1-\alpha)100\% \text{ Confidence Level.}$$

Show that if two variables are independent, their correlation will be zero and their covariance will be zero.

Definition of correlation between two variables, X and Y:

$$Corr(X,Y) = \frac{Cov(X,Y)}{\sigma_X \sigma_Y} = \frac{E[(X - E[X])(Y - E[Y])]}{\sigma_X \sigma_Y} = \frac{E[(X - \mu_X)(Y - \mu_Y)]}{\sigma_X \sigma_Y}$$
$$= \frac{E[XY - X\mu_Y - Y\mu_X + \mu_X\mu_Y]}{\sigma_X \sigma_Y} = \frac{E[XY] - E[X]\mu_Y - E[Y]\mu_X + E[\mu_X\mu_Y]}{\sigma_X \sigma_Y}$$
$$= \frac{E[XY] - \mu_X\mu_Y - \mu_Y\mu_X + \mu_X\mu_Y}{\sigma_X \sigma_Y} = \frac{E[XY] - \mu_Y\mu_X}{\sigma_X \sigma_Y}.$$

Definition of independence of two random variables, X and Y:

If X and Y are independent, and $f(x,y)$ is their joint pdf, then there exists $g(x)$ and $h(y)$, the pdf of X and Y, respectively, such that $f(x,y) = g(x)h(y)$. As a consequence
$$E[XY] = \iint xy\, f(x,y) = \iint xy\, g(x)h(y)dxdy = \int x\, g(x)dx \int yh(y)dy$$
$$= E[X]E[Y], \text{ so } Corr(X,Y) = \frac{E[XY] - \mu_Y\mu_X}{\sigma_X \sigma_Y} = \frac{E[X]E[Y] - \mu_Y\mu_X}{\sigma_X \sigma_Y}$$
$$= \frac{\mu_Y\mu_X - \mu_Y\mu_X}{\sigma_X \sigma_Y} = 0$$

Therefore, if two variables, X and Y are independent, $Corr(X,Y) = 0$ and $Cov(X,Y) = 0$.

Calculate Pearson's correlation coefficient for two linearly dependent variables

Pearson's correlation coefficient is the covariance of two random variables divided by the product of their standard deviations.
$$Corr(X,Y) = \frac{Cov(X,Y)}{\sigma_X \sigma_Y} = \frac{E[(X - E[X])(Y - E[Y])]}{\sigma_X \sigma_Y}$$
$$= \frac{E[(X - \mu_X)(Y - \mu_Y)]}{\sigma_X \sigma_Y} = \frac{E[XY - X\mu_Y - Y\mu_X + \mu_X\mu_Y]}{\sigma_X \sigma_Y}.$$
It is standardized in the sense that it is equal to unity when $(X,Y) = \sigma_X \sigma_Y$. It is non-zero when $Cov(X,Y) \neq 0$.

Since $Cov(X,Y) = E[XY - X\mu_Y - Y\mu_X + \mu_X\mu_Y]$, it can only detect linear correlations between random variables X and Y. Note: Linear, in this case, is defined as a function between x and y where all the exponents of the variables are unity. Its value ranges from -1 to 1.

Example #1: If $y = x$, y is linearly dependent on x, and $Corr(X,Y) = 1$,
$$Corr(X,Y) = \frac{E[(X - E[X])(Y - E[Y])]}{\sigma_X \sigma_Y} = \frac{E[(X - \mu_X)(X - \mu_X)]}{\sigma_X \sigma_X} = \frac{Var(X)}{Var(X)} = 1.$$

Note: Recall the definition $Var(X) = E[(X - E[X])^2] = E[(X - \mu_X)^2] = E[(X - \mu_X)(X - \mu X.$
Example #2: If $y = -x$, y is linearly dependent on x, and $Corr(X,Y) = -1$,
$$Corr(X,Y) = \frac{E[(X - E[X])(Y - E[Y])]}{\sigma_X \sigma_Y} = \frac{[(X - E[X])(-X - E[-X])]}{\sigma_X \sigma_Y}$$
$= \frac{E[(X-\mu_X)(-X+\mu_X)]}{\sigma_X \sigma_X} = \frac{-Var(X)}{Var(X)} = -1$. Pearson's Correlation Coefficient is often used to estimate linear relationships.

Show that if the Pearson correlation coefficient of two variables is zero, it does not imply the variables are independent, particularly if the variables have a non-linear correlation

If $Corr(X, Y) = 0$, it implies that $Cov(X, Y) = 0$ and the slope of the least-squares regression line will also be zero. Nevertheless, we will now present an example in which a non-linear dependence between X and Y has $Corr(X, Y) = 0$ and $Cov(X, Y) = 0$.

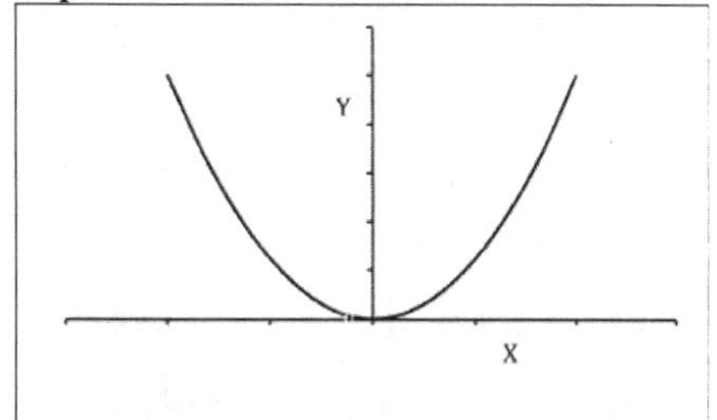

Example #2: If $y = x^2$, y is dependent on x.
$E(X) = \mu_X = 0$ because for every X equal to a positive number, there is a number of the same magnitude but opposite sign. So,
$E[(X - \mu_X)(Y - \mu_Y)] = E[(X)(Y - \mu_Y)]$.
For every value of Y, there are two values of X that are equal in magnitude and of opposite sign. Therefore, $E[(X)(Y - \mu_Y)] = 0$, and

$$Corr(X, Y) = \frac{Cov(X, Y)}{\sigma_X \sigma_Y} = \frac{E[(X - \mu_X)(Y - \mu_Y)]}{\sigma_X \sigma_Y} = 0.$$

Therefore, $Corr(X, Y) = 0$ does not imply X and Y are independent.
Note: $Corr(X, Y)$ can only be relied upon to detect linear correlations between variables.

How a linear least-square regression line is computed from a set of data $\{(x_i, y_i)\}$, and define its slope in terms of $Cov(x, y)$ and $Var(x)$

A set of data $\{(x_i, y_i)\}$ is plotted. The linear regression line is $\hat{y}_i = A + Bx_i$, where $\hat{y}_i$ is the value predicted for y_i in the data pair (x_i, y_i).

The error is,
$y_i - \hat{y}_i = y_i - A - Bx_i$.
The sum of the square of the errors for all data points is

$$\sum_{i=1}^{i=n} (y_i - A - Bx_i)^2.$$

A and B are adjusted until this sum is minimized,
$Intercet = A = \bar{y}_n - B\bar{x}_n$

$$Slope = B = \frac{Cov(X, Y)}{Var(X)} = r\sqrt{\frac{Var(Y)}{Var(X)}},$$

where $Cov(x, y)$ is the covariance of x and y, and r is known as the correlation coefficient.

Practice Test

Practice Questions

1. Students at Sunnyside High School participate in football and band. 35% of students play football and play in the band. 42% of students play football. Approximately what percentage of students who play football also march in the band?
   a. 7%
   b. 15%
   c. 23%
   d. 83%

2. There are five favorite parking spaces in the game, Shopping Extravaganza. You receive 2 tokens if you land on Parking Space #1, 6 tokens if you land on Parking Space #2, 2 tokens if you land on Parking Space #3, 3 tokens if you land on Parking Space #4, and 7 tokens if you land on Parking Space #5. You have an equal chance of landing on each space. What is the expected value?
   a. 4
   b. 7
   c. 10
   d. 20

3. In which scenario would stratified random sampling be the best sampling choice?
   a. You want to examine the differences in student performance on a college entrance exam across high school districts.
   b. You want to examine preferences in coffee shops, according to college students, work professional, social groups, and other categories of coffee drinkers.
   c. You want to examine student attitude towards mathematics, broken down by gender.
   d. You want to examine vacation preferences of all co-workers.

4. Which of the following contributes to survey bias?
   I. Use of a convenience sample
   II. Use of a small sample size
   III. Inclusion of voluntary responses
   IV. Use of phone calls or mailed surveys
   a. I only
   b. II and III
   c. I, III, and IV
   d. I, II, III, and IV

5. Suppose I receive a certain amount of marbles when the arrow on an eight-section spinner lands on various numbers. I receive 2 marbles each time the arrow lands on a 2 or 3, 4 marbles each time the arrow lands on the 6, 8 marbles each time the arrow lands on a 7 or 8, and no marbles when the arrow lands on any of the other numbers. What is the expected value for the number of marbles I will receive?

a. 3
b. 5
c. 15
d. 24

6. Look at the table. The table shows the coffee drink preferences of mathematics faculty.

	Latte	Frappuccino	Cappuccino	Iced Coffee	Total
TAMU	23	36	19	8	86
ASU	19	29	26	5	79
Total	42	65	45	13	165

What is the approximate probability of TAMU faculty or cappuccino drinker?

a. 12%
b. 42%
c. 68%
d. 79%

7. Northern Arizona University integrates a new software into its introductory statistics course. With a sample size of 35 students, the sample mean is 5.6 points higher per student with a sample standard deviation equal to 1.2. Use $\alpha = .05$. What can be concluded? Describe your process.

8. Results from the National Association of State Foresters reveal an average profit of \$120/acre per year with $\sigma = \$24$. From a survey of 15 state forestry associations, the average profit was reported to be \$126/acre per year. Compute a 95% confidence interval and interpret the results.

9. Puffmakers Marshmallow Company lists 16 ounces on each bag. Suppose $\sigma = 1.3$ ounces. You purchase 34 bags and find $\bar{X} = 15.1$. Is this finding rare? Explain why or why not. $\propto = .05$ should be used to determine the cutoff.

10. A new drug for treating Alzheimer's disease has caused side effects in 5% of the people taking the drug. In the state of New York, 250 people are sampled, and 19 people are found to have side effects. What do these results say about the sample? Use .05 as the critical probability.

11. Sally's deli makes an average monthly profit of \$14,500 with a standard deviation of \$1,200. In June, she made \$16,800. In July, she made \$13,700. How many standard deviations above and below her monthly profit was she during the months of June and July?

a. 4.67, −12.08
b. 1.92, −.67
c. 2.58, −3.26
d. 67, −1.92

12. Given the monthly car payments (in dollars) made in Forks, Washington during 2009:
290 315 345 520 620 435 210 560 380 475 615 645 205 425
570 630 235 490 285 640 380 525 635 465 205 445 505 485
Create a histogram and describe the process used. Interpret the skewness and kurtosis.

13. Which university had the most variation in tuition cost over the past five years?

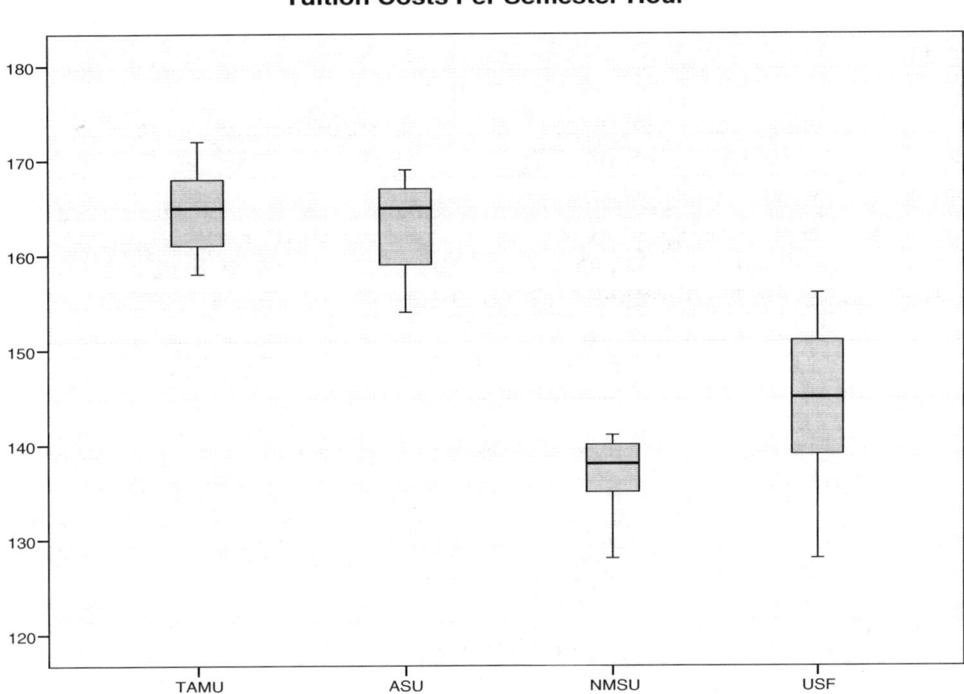

a. TAMU
b. ASU
c. NMSU
d. USF

14. The manager of a local gym, called Work-N-Fit, wishes to increase the personal trainer charge. He conducted a survey of Work-N-Fit gym members to determine their reaction. What is the population?

a. All area gym members
b. All Work-N-Fit gym members who have a personal trainer
c. All Work-N-Fit gym members
d. All area gym members who participate in a personal trainer program

15. Which scenario describes an experiment?

a. The census bureau analyzes the 2010 household data
b. A school principal analyzes the effects of a new curriculum on standardized test scores
c. A sociologist observes the behaviors of couples at a social function
d. Rasmussen analyzes results to various political polls

16. Which of the following, concerning a normal distribution, is *true*?

I. Approximately 97% of the area under the curve is within 3 standard deviations of the mean

II.A normal distribution has a standard deviation of 1

III.A normal distribution has a mean of 0

IV. The area is 1

a. I and II
b. II, III, and IV
c. I, II, III, and IV
d. I, II, and III

17. If we sample 45 graduates and then increase our sample size to 90 graduates, what will happen to the shape of the overall distribution?

a. The distribution will not change
b. The distribution will become more peaked and thus show less variance
c. The distribution will become less peaked and thus show more variance
d. There isn't enough information given to answer the problem

18. Suppose the null hypothesis states that House Bill 26351 passes. You fail to reject the null hypothesis, when the null hypothesis was indeed false. What type of error have you committed?

a. Type I error
b. Type II error
c. Neither
d. There isn't enough information given to answer the problem

19. Which of the following is *true*, regarding power of a test?

a. Type I error decreases as more samples are included
b. The power is the probability of not making a Type I error
c. Power is calculated by subtracting the probability of a Type II error from alpha
d. The power is the probability of not making a Type II error

20. Which of the following is the sample size needed to produce a sample mean with margin of error of 25%? The standard deviation is 2 and the confidence level is 95%.

a. 48
b. 61
c. 173
d. 246

21. Elizabeth desires to be in the 90th percentile of all students taking the verbal portion of the MCAT exam. $\mu = 11, \sigma = 1.5$. Approximately what score must she make to be in the 90th percentile?

a. 11
b. 12
c. 13
d. 14

22. Jana must draw a name out of a hat. There are 15 names in the hat. On the first try, she draws "Edward", which is her brother. She must replace the name. What is the probability that she will draw out the name "Edward" again?

a. $\frac{1}{15}$
b. $\frac{1}{225}$
c. 15
d. 225

23. A medical researcher wants to determine the correlation between age of patients with a fungal disease and severity level of the disease, categorized on a scale of 1 – 20. Plot the given data in a scatter plot, draw the least squares regression line, and find the equation for the least squares regression line.

Age	Severity
25	4
45	16
40	11
55	19
28	8
29	7
31	8
21	2
25	5
32	8
41	14
44	14
45	15
41	12
28	6
29	7
36	10
38	11
41	13
51	18

24. A top vehicle manufacturing company decides to increase vehicle costs on all models by 3 percent to balance the growing overhead costs. The current standard deviation for vehicle prices is $2000. What will be the standard deviation for the new vehicle prices?

a. 60
b. 600
c. 2060
d. 2600

25. The percentages of students in College Station, Texas, using each bank during 2008 is shown below:

Bank	**Compass Bank**	**Wells Fargo**	**Commerce National**	**Bank of America**	**Other**
Percentage	.27	.31	.14	.24	.04

The accounting division at Texas A&M University wants to determine whether or not these findings are consistent with the findings from 2009. A random sample of 200 students reveals the following data:

Bank	**Compass Bank**	**Wells Fargo**	**Commerce National**	**Bank of America**	**Other**
Number of Students	70	88	21	15	6

Describe whether or not the distribution is the same as last year. Show the process. Use .05 as the level of significance.

26. The principal at Lake Mary High School wants to determine the effects of a new mathematics curriculum on student's final scores. She wishes to compare the new curriculum with two sets of old curriculum. Students taught by Teacher A, Teacher B, and Teacher C will be used in the sample. The students taught by each teacher are exposed to different teaching methods. Which design would be the best choice when comparing the effects of the three curricula?

a. Completely randomized design
b. Randomized block design
c. Matched pairs design
d. Replicated design

27. Provide an example to illustrate the Law of Large Numbers. Explain each part of the example. Be certain to state the meaning of the law within the explanation.

28. Dr. Thomas wishes to determine if grade average from EPSY 640 is related to grade average when taking EPSY 641. He selects a random sample of 10 students who have taken both classes. The data is shown below.

Student	EPSY 640 Grade	EPSY 641 Grade
Student 1	84	87
Student 2	79	91
Student 3	94	91
Student 4	86	80
Student 5	68	69
Student 6	75	70
Student 7	82	88
Student 8	87	90
Student 9	90	82
Student 10	79	85

Which of the following is the most approximate correlation coefficient for the data?

a. 61
b. 65
c. 68
d. 71

29. A traveling sales consultant drives to potential customers' houses on a daily basis. The following shows the miles driven per day over a 20 day period:

	Monday	Tuesday	Wednesday	Thursday	Friday
Week 1	45	32	51	64	38
Week 2	41	19	49	27	35
Week 3	56	43	26	17	61
Week 4	51	22	36	44	37

On 15 percent of the days, what is the approximate *minimum* number of miles driven per day?

a. 22
b. 36
c. 51
d. 56

30. The table shows movie genre preferences according to undergraduate and graduate students:

	Action	Comedy	Drama	Romance	Totals
Undergraduate Students	42	51	36	28	157
Graduate Students	32	44	41	25	142
Totals	74	95	77	53	299

What is the approximate expected number of graduate students with a drama preference?

a. 11
b. 19
c. 37
d. 41

31. The prices for a gallon of milk across New Mexico are normally distributed, with $\mu = \$2.99$ and $\sigma = \$.75$. The Bureau of Labor Statistics samples 65 grocery stores in the state of New Mexico. What is the probability that the prices for a gallon of milk are between $2.85 and $3.25?

a. ≈ 84%
b. ≈ 87%
c. ≈ 95%
d. ≈ 99%

32. The variance for the income tax return value for all households in Oro Valley, Arizona is unknown. A random sample of 12 households from Oro Valley, Arizona reveals the following data:

Household	Income Tax Return
Household A	$120
Household B	$2,566
Household C	$345
Household D	$2,166
Household E	$892
Household F	$0
Household G	$26
Household H	$564
Household I	$1,034
Household J	$410
Household K	$0
Household L	$219

What is the approximate point estimate for the *variance*?

a. $853
b. $40,272
c.$667,492
d. 728,174

33. The CEO of a top vehicle brand randomly selects 150 cars from all cars produced at a particular manufacturing facility. Of the 150 cars, 26 of them are found to have a defect. What is the true proportion of defective cars at this facility? Compute a 95% confidence interval.

a. Between .11 and .17
b. Between .11 and .23
c. Between .15 and .19
d. Between .17 and .22

34. Texas A&M University is a large and prestigious university in the state of Texas. An impartial research professional is solicited to determine and compare the proportion of high school students that plan to attend the university during the next fall semester. The research professional randomly selects 80 students from Wichita Falls High School and finds 45 students who plan to attend the university. He randomly selects 75 students from Corpus Christi High School and finds 39 students who plan to attend the university. Which of the following is the best estimate for the difference in proportion of students who plan to attend the university, as gathered from the two high schools? Compute a 90% confidence interval.

a. Between −.09 and .17
b. Between −.02 and .20
c. Between .39 and .65
d. Between .43 and .69

35. A medical doctor wishes to examine the effects of a new drug, used to treat insomnia. He randomly selects 16 of his patients to be participants in the study and randomly places the participants in two different groups. One group is selected at random, whereby the group receives 2 weeks of treatment with the drug and another 2 weeks with a placebo. A card is drawn out of a hat to determine which weeks to provide the treatment drug. The average number of hours slept per night is recorded.

Number of Hours Slept Per Night

Participant	Treatment Drug	Placebo
Participant A	5	4
Participant B	8	5
Participant C	7	2
Participant D	8	5
Participant E	6	5
Participant F	9	7
Participant G	6	6
Participant H	7	3
Participant I	6	4
Participant J	8	3
Participant K	7	6
Participant L	7	5
Participant M	7	4
Participant N	7	6
Participant O	7	4
Participant P	5	3

Using a 95% confidence interval, what is the true mean difference in number of hours slept per night?

a. Between .68 and 2.22
b. Between .81 and 2.09
c. Between 1.61 and 3.15
d. Between 1.74 and 3.01

36. The New York State Education Commissioner examines the number of students passing the mathematics portion of the 2010 Regents Exam. A total of 30 randomly selected schools, with 40 randomly selected students per school, are used in the sample. The number of students not passing is as follows:

# of students not passing	0	1	2	3	4
# of schools	6	8	4	3	9

Show whether or not the binomial distribution should be used to model the data.

37. Provide a scenario when you would need to use a matched pairs design. Explain the technique and steps taken to set up such a design for the given scenario.

38. The president of a large, mathematics software company wishes to determine the interquartile range in weekday sales over the past month.

$150	$220	$310	$100	$140
$380	$290	$160	$190	$90
$420	$180	$130	$120	$160
$290	$240	$180	$210	$180

Which of the following is the correct interquartile range?

a. 85
b. 120
c. 140
d. 150

39. The following cumulative frequency chart shows the maximum amount of money (in dollars) spent on coffee per month by students.

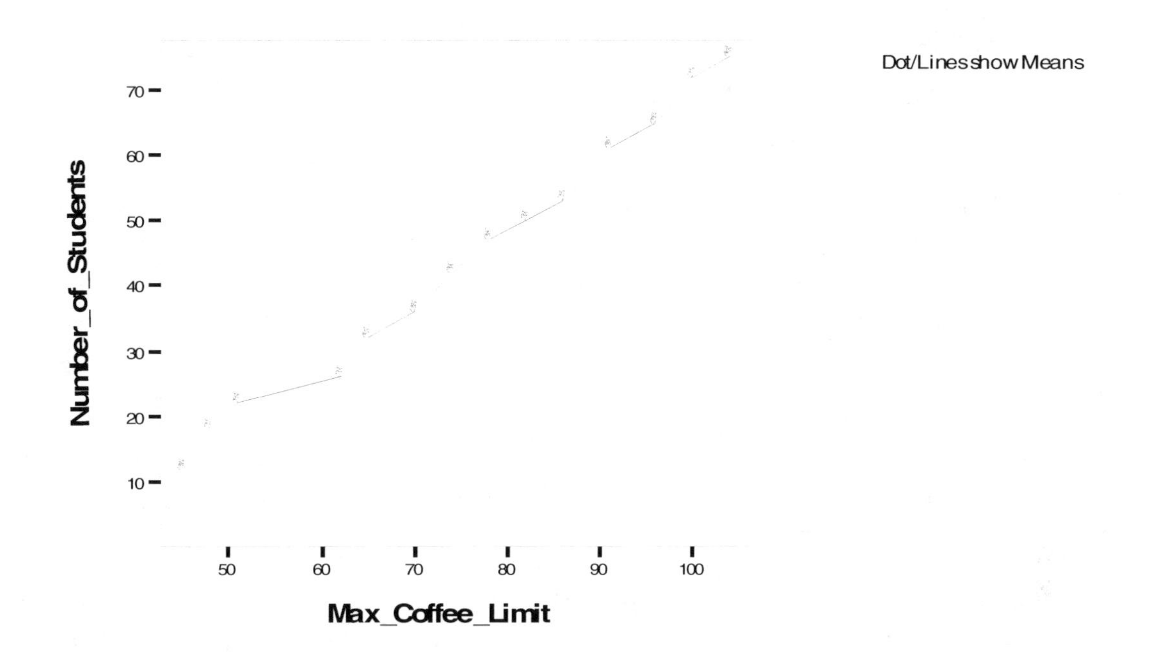

Which of the following is the best estimate for the number of students that spend less than $70 per month on coffee?

a. 33
b. 75
c. 106
d. 139

40. An amateur golfer claims an average putting score on a 18-hole course to be 97. Suppose the standard deviation is 2.5. On 35 different tournament days, $\bar{X}$=102. Determine if there is a significant difference between the two means. Explain your reasoning.

41. A popular fast food chain is giving away a certain number of winning codes in each value meal. Of the 1,216 randomly selected customers surveyed in Charlotte, North Carolina, it was determined that during one week, 4% of the customers received a winning code. Of the 1,310 randomly selected customers surveyed in Flint, Michigan, it was determined that during the same week, 6% of the customers received a winning code. Determine which of the following is the correct z-score for the difference in proportions and determine whether or not the difference is significant.

a. −1.8, no significant difference
b. −1.8, significant difference
c. −2.3, no significant difference
d. −2.3, significant difference

42. The table below shows the SAT and GMAT scores of 12 randomly selected University of Montana students.

SAT Scores	GMAT Scores
820	450
1260	600
2100	780
960	480
680	260
1450	620
1310	580
1020	360
1680	410
2060	710
1810	690
1430	440

Compute the predicted GMAT scores and residual values. Make a residual plot of the residual values and predicted values.

43. A scatterplot is used to analyze the correlation between average cost of home and average salary in Leesburg, Virginia. The x-values, or values for the average cost of home, are rather large. It is desired that the effect of the large values of x be reduced. Which transformation would be most appropriate to obtain linearity?

a. Log transformation
b. Square root transformation
c. Reciprocal transformation
d. Square transformation

44. A health insurance company sends trained sales personnel all across the state of Idaho to solicit new workers. The traveling sales personnel are divided into two regional groups: Region A and Region B. The table below shows the number of miles traveled per day by personnel in each group.

Region A	**Region B**
53, 82, 17, 90, 67,	81, 10, 29, 73, 31,
72, 29, 92, 75, 22,	64, 86, 11, 48, 68,
86, 91, 34, 73, 89,	29, 35, 42, 40, 52,
95, 41, 50, 62, 71,	23, 18, 30, 43, 59,
86, 43, 56, 64, 69,	51, 16, 26, 35, 56,
72, 51, 83, 94, 68	36, 42, 19, 34, 23

Create a back-to-back stemplot and describe the overall shape of the distribution for each region. Relate the shape of the distribution to the context of the problem.

45. A new hybrid vehicle from one of the top manufacturers claims gas mileage of 34 miles per gallon. In 30 randomly selected cars, the following mileage data per gallon is gathered:

33.2	34.5	33.8	34.2	35.1	36.2	32.9	33.8	34.4	35.4
33.6	34.6	32.8	36.1	34.9	33.5	33.7	34.8	33.3	33.7
32.7	33.0	34.1	32.9	31.8	34.4	36.1	35.5	33.6	32.5

What is the true mean gas mileage for the new hybrid vehicle? Use a 90 percent confidence level.

a. Between 34.77 and 35.31
b. Between 33.69 and 34.39
c. Between 32.86 and 33.17
d. Between 31.54 and 32.06

46. A top executive at a large Informational Technology company is up for tenure. Employees throughout the building are asked whether or not they believe the executive should be promoted. The response choices are simply "yes" or "no". Employees are divided into four categories: Intern, Entry-Level, Mid-Level, and Senior Level. A random sample of 35 employees from each of the two lower level categories and a random sample of 45 employees from each of the two upper level categories is taken. The table below summarizes the findings:

	Yes	**No**	*n*
Intern	24	11	35
Entry-Level	29	6	35
Mid-Level	38	7	45
Senior Level	31	14	45

Determine whether or not there is homogeneity across the proportions. Show the test used and explain the process.

47. When performing a t-test for the slope of a least squares regression line, which of the following is *not* a required condition?

a. A random sample of pairs is used
b. The residuals appear to be normally distributed
c. The standard deviation of the residuals is 1
d. The residual plot does not show a particular pattern

48. A high school principal wants to investigate the effects of "Clickers in the Classroom" on students' mathematics achievement. Students are randomly placed in two groups, with one group receiving clickers and the other group not receiving any clickers. Student achievement is determined by scores on the latest mathematics unit test. The group receiving clickers, with $n = 45$, reported a group average of 87.2, with a group variance of 43.8. The group not receiving clickers, with n = 48, reported a group average of 85.6, with a variance of 45.9. Use $\alpha = .05$. Which of the following shows the correct observed t-value and decision regarding the effect of clickers on mathematics achievement?

a. t = .23; no significant difference in achievement
b. t = 1.15; no significant difference in achievement
c. t = 1.39; significant difference in achievement
d. t = 6.7; significant difference in achievement

49. Robert sells 45 vacuum cleaners this month. He was given 70 vacuum cleaners to sell. In order to be awarded a bonus, he must sell 75% of the vacuum cleaners each month. Suppose the level of significance is .05. Determine whether you will reject or fail to reject the claim that he sells 75% of the 70 vacuum cleaners. Explain your reasoning.

50. The university Student Government Association wishes to determine the proportion of students that utilize the new cyber café. Of the 450 students interviewed, 321 report that they do utilize the cyber café. Which of the following is the standard deviation of the sample proportion?

a. 02
b. 12
c. 41
d. 2.45

Answers and Explanations

1. D:

$P(\text{B given A}) = \frac{P(\text{A and B})}{P(\text{A})}$

$= \frac{.35}{.42}$

$= .833$

$\approx 83\%$

Approximately 83% of students.

2. A:

$E(X) = \left(2 \cdot \frac{1}{5}\right) + \left(6 \cdot \frac{1}{5}\right) + \left(2 \cdot \frac{1}{5}\right) + \left(3 \cdot \frac{1}{5}\right) + \left(7 \cdot \frac{1}{5}\right)$

$= \frac{2}{5} + \frac{6}{5} + \frac{2}{5} + \frac{3}{5} + \frac{7}{5}$

$= \frac{20}{5}$

$= 4$

3. C: Stratified random sampling allows the individual to break a sample into groups, or strata. Then, a random sample can be taken from each group. If you want to investigate differences in attitude, according to gender, a stratified random sampling technique would be best.

Choice A is not correct because you typically want to investigate the overall performance per high school in such a study.

Choice B is very similar to Choice C. However, Choice B lends itself more to cluster sampling. With cluster sampling, the sample is broken into categories or groups. All participants are used within each category. Cluster sampling is typically used when several categories are to be examined.

Choice D is not correct simply because you are looking at all co-workers, which would serve as a population, not a sample

4. C: All of the items are sources of survey bias, except for usage of a small sample size. Random sampling is the key to avoiding sampling bias, not sample size. A biased sample can have a large sample size, just as an unbiased sample can have a small sample size.

5. A:

$E(X) = \left(2 \cdot \frac{1}{8}\right) + \left(2 \cdot \frac{1}{8}\right) + \left(4 \cdot \frac{1}{8}\right) + \left(8 \cdot \frac{1}{8}\right) + \left(8 \cdot \frac{1}{8}\right) + \left(0 \cdot \frac{1}{8}\right) + \left(0 \cdot \frac{1}{8}\right) + \left(0 \cdot \frac{1}{8}\right)$

$= \frac{2}{8} + \frac{2}{8} + \frac{4}{8} + \frac{8}{8} + \frac{8}{8} + 0 + 0 + 0$

$= \frac{24}{8}$

$= 3$

The expected value is 3 marbles.

6. C: P(TAMU or Cappuccino) = P(TAMU) + P(Cappuccino) − P(TAMU and Cappuccino)

$= \frac{86}{165} + \frac{45}{165} - \frac{19}{165}$

$= \frac{112}{165}$

$\approx 68\%$

7. The new software significantly increases student achievement in the stats course.
Manual calculation of the observed t-value gives:

$$t = \frac{|x_1 - x_2|}{\sqrt{\frac{s_1^2}{n_1} + \frac{s_2^2}{n_2}}}$$

$$= \frac{5.6}{\sqrt{\frac{(1.2)^2}{35}}}$$

$$= 27.61$$

For df of 34, the critical t-value is 1.684. Since 27.61 is greater than 1.684, we will reject the null hypothesis, thus concluding there is a difference.
We can also use the graphing calculator to find the p-value.
Assuming:
Null hypothesis: $\mu = 0$
Alternate hypothesis: $\mu > 0$
We will run a T-test.
Go to STAT
Scroll to Tests and choose T-test
Choose Input of Stats
Enter $\mu_0 = 0$, $\overline{X} = 5.6, s_x = 1.2$, and $n = 35$
Choose > μ_0
Choose Calculate

Find the p-value. Since the p-value is less than .05, we will reject the null hypothesis. (Notice we get the same t-value as above.)
Thus, the new software significantly increases student achievement in the stats course.

8. Manually compute the confidence interval:

$$\bar{X} \pm z_{.025} \frac{\sigma}{\sqrt{n}}$$

$$= 126 \pm 1.96 \frac{24}{\sqrt{15}}$$

$$= 126 \pm 12.15$$

$$= 113.85 \text{ or } 138.15$$

Since 120 is in the interval, the average profit for the 15 state forestry associations is consistent with the average profit for the National Association of State Foresters.
You can also compute the confidence interval on a graphing calculator.
Go to STAT
Scroll over to Tests and choose Z-Interval
Choose Input Stats
Enter $\sigma = 24, \bar{X} = 126$, and $n = 15$
Keep the default of C-Level of .95
Choose Calculate
The 95% confidence interval = 113.85 to 138.15.

9. Manual calculation of the observed z-value gives:

$$Z = \frac{\bar{X}-\mu}{\sigma/\sqrt{n}}$$

$$= \frac{15.1-16}{1.3/\sqrt{34}}$$

$$= \frac{-.9}{.2229}$$

$$= -4.04$$

The z-value results in an area less than .05. The probability is actually 0. Thus, we will reject the null hypothesis and conclude that the findings are rare.
We can also use the graphing calculator to find the p-value.
Go to STATS
Scroll over to Tests and choose Z-Test
Choose Input Stats
Enter $\mu = 16, \sigma = 1.3, \bar{X} = 15.1, n = 34$
Choose $< \mu_0$
Calculate
The p-value or area is 2.71 E-5. Since the p-value is less than .05, we will reject the null hypothesis and conclude the findings are rare.

10. Solve using a binomial probability
$P(X \geq 19) = \sum_{x=19}^{250} \binom{250}{x} \cdot (.05)^x \cdot (.95)^{250-x}$
You can also solve the problem using the following:
$P(X \geq 19) = 1 - P(X \leq 18)$
$= (1 - \text{binomcdf}(250, .05, 18))$
$= .047$

The probability is .047. Since .047 is less than .05, it is "rare" or "unusual" for 19 people out of 250 people to have experienced side effects.

11. B: Compute a z-score for June.

$$\text{z-score} = \frac{\text{June profit} - \text{average monthly profit}}{\text{standard deviation}}$$

$$= \frac{16{,}800-14{,}500}{1{,}200}$$

$$= 1.92$$

Her June profit was 1.92 standard deviations above the average monthly profit.
Compute a z-score for July.

$$\text{z-score} = \frac{\text{July profit} - \text{average monthly profit}}{\text{standard deviation}}$$

$$= \frac{13{,}700-14{,}500}{1{,}200}$$

$$= -.67$$

Her July profit was .67 standard deviations below the average monthly profit.

12. The created histogram should look like:

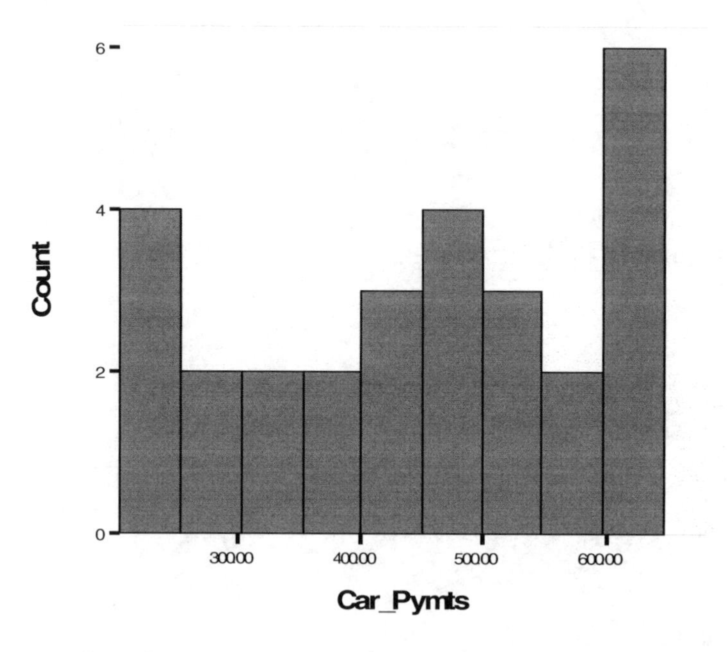

An overlay of the normal curve reveals:

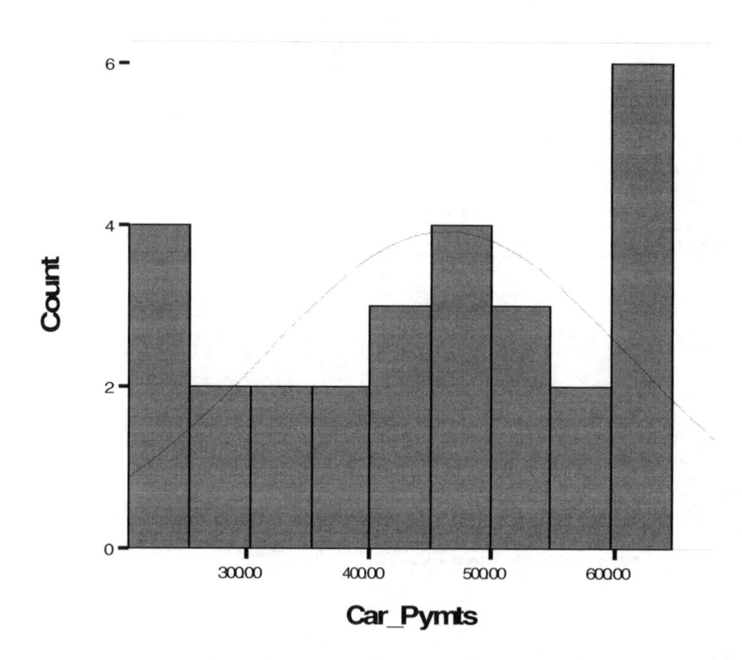

Skewness relates to the symmetry. Since the histogram is skewed left, we know it has negative skewness. Kurtosis describes the sharpness of the peak. Since the distribution is flatter than the normal distribution, we have a negative kurtosis. Thus, we have more variation in our data. In summary, the car payments are skewed in the direction of $450 or less per month. We also might state that the negative kurtosis, i.e., flatter peak, reveals much variation in car payments in Forks, Washington.

13. D: The University of South Florida (USF) shows the most variation between costs. This box plot has the greatest interquartile range, as well as greatest overall range.

14. C: The population is all Work-N-Fit gym members because the increase of personal trainer charge will only apply to those gym members.
Choice A is not correct because all area gym members will not be involved in or affected by the decision.
Choice B is not correct because the gym members do not need to have a personal trainer to be part of the population.
Choice D is not correct because gym members outside of Work-N-Fit, regardless of their participation in a personal trainer program, are not involved in or affected by the decision.

15. B: The investigation of the effects of an intervention is indeed an experiment.
Choice A is not correct because no treatment is given.
Choice C is not correct because only observation takes place, with no treatment given.
Choice D is not correct because no treatment is given.

16. B: A normal distribution has a standard deviation of 1, a mean of 0, and an area of 1. There is approximately 99.7% of area under the curve within 3 standard deviations of the mean, not 97%.

17. B: This problem relates to the Central Limit Theorem. As the sample size increases, the sample mean approaches the population mean. In other words, as the sample size increases, the distribution approaches the normal distribution. Thus, there is less variance, with a more peaked distribution.

18. B: When you fail to reject a null hypothesis that is indeed false, you have committed a Type II error.
Choice A is not correct because a Type I error is only committed when you reject a null hypothesis that is true.

19. D: Power is the probability of not making a Type II error. In other words, power is the probability of not failing to reject a null hypothesis that is indeed false.
Choice A is not correct because sample size cannot decrease Type I error. Sample size can only increase the possibility of finding significance.
Choice B is not correct because power is the probability of not making a Type II error, not a Type I error.
Choice C is not correct because power is actually calculated by subtracting the probability of a Type II error from 1, not alpha.

20. D: In order to determine the sample size needed to produce a sample mean with a given margin of error, standard deviation, and desired confidence level, you use the formula below:

$$\left(\frac{Z_{\alpha/2} \cdot \sigma}{E}\right)^2 = \left(\frac{1.96 \cdot 2}{.25}\right)^2 = (15.68)^2 = 245.86$$

Since you cannot have a fraction of a person, you must round up to the next whole number, which is 246 in this case. Thus, you need a sample size of 246 to achieve the desired margin of error with given standard deviation and confidence level.

21. C: In this situation, you should have a normal distribution of the form
X ~ N(11, 1.5)
Using the invNorm function of the graphing calculator, you will plug in the given values into the form:
invNorm (%, μ, σ)
Thus, we enter invNorm (.90, 11, 1.5), getting an answer of 12.92. This score rounds to 13.

22. A: This situation describes a compound event that involves two independent events. Thus, P (A + B) = P(A) × P(B). The sample space is 15. Thus, we have $\frac{1}{15} \times \frac{1}{15} = \frac{1}{225}$.

23. The scatterplot (with least squares regression line) should look like:

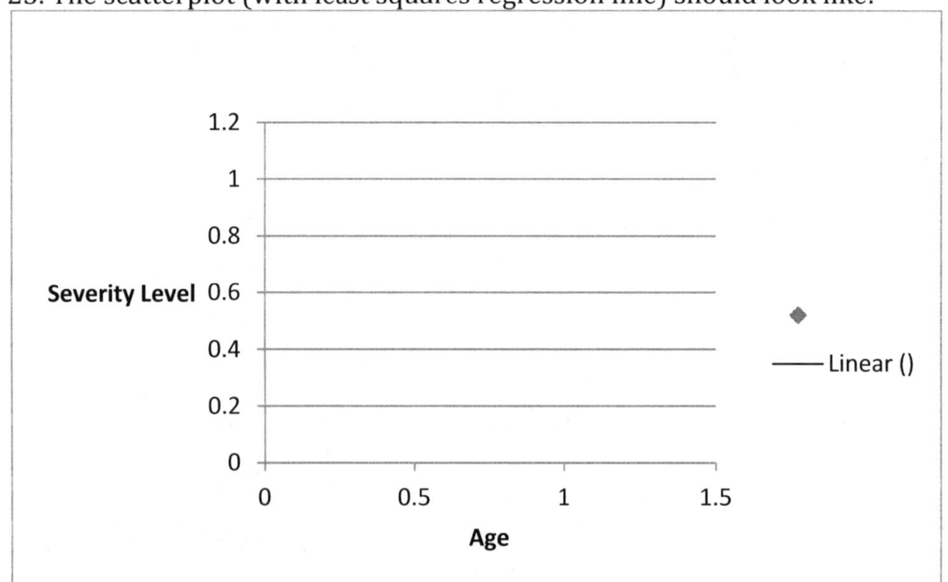

The least squares regression equation that models the correlation is $y = -7.73 + .5x$.
Using the calculator, you simply enter the x-values into L1 and enter the y-values into L2. Then, go to Stat, Calculate, and lastly LinReg (a + bx). Enter L1,L2 and click "enter". Doing so will give you the a and b values you need.
You can also calculate the least squares regression equation manually. First, calculate the sum of squares for x and sum of squares for y (27,921 and 2584, respectively). Then find $\bar{X}$ and $\bar{Y}$. The sample mean of x ($\bar{X}$) can be found by summing the x-values and dividing by n:
$\frac{725}{20} = 36.25$
The sample mean of y ($\bar{Y}$) can be found by summing the y-values and dividing by n:
$\frac{208}{20} = 10.4$
We will now calculate the slope.
$b = \frac{SS_{xy}}{SS_{xx}}$
We must first find each part.
S_{xx} is found by starting with the raw sum of squares for x and subtracting the quantity (n multiplied by ($\bar{X}^2$)). In other words, you have:
$S_{xx} = 27921 - 20(1314.06) = 1639.8$
S_{xy} is found by summing the products of $(X - \bar{X})(Y - \bar{Y})$. $S_{xy} = 796$
Finally,
$b = \frac{SS_{xy}}{SS_{xx}} = \frac{796}{1639.8} = .49$

We will now calculate the y-intercept.
$a = \bar{Y} - b\bar{X}$
Thus we have:
$a = 10.4 - .49(36.25) = -7.36$
Putting the slope and y-intercept into the equation gives:
$y = -7.36 + .49x$
This answer is slightly different from the answer given by the graphing calculator, simply because of rounding.

24. C: Since each buyer will pay 3 percent more per vehicle, the new vehicle price will be:
new vehicle price = 1.03(old vehicle price)
To find the new standard deviation, you simply multiply the absolute value of the new vehicle price by the old standard deviation. Thus,
$|1.03|(2000) = 2060$

25. The results for percentage of students using each bank in 2009 are different from the results recorded in 2008.
Since students are classified according to categories of banks, and we are given percentages of each, we can use a chi-square goodness of fit test.
Let's suppose the following:
Null hypothesis: The percentage of students using each bank in 2009 is the same as in 2008.
Alternative hypothesis: The percentage of students using each bank in 2009 is different from the percentage in 2008.
We need to compare apples to apples by comparing the actual number of students using each bank in 2009 to the expected number of students using each bank in 2009, based upon the percentage data we were given for 2008. Therefore, we will take our sample size and multiply it by the 2008 percentage for each bank.

Bank	**Compass Bank**	**Wells Fargo**	**Commerce National**	**Bank of America**	**Other**
Observed Sample in 2009 (Number of Students)	70	88	21	15	6
Expected Number of Students	54	62	28	48	8

Our *df* is equal to 4 (k categories minus 1); since all expected values are greater than 4, we are fine.
We were told to use a .05 level of significance. The chi-squared formula is:
$x^2 = \sum_{i=1}^{5} \frac{(O_i - E_i)}{E_i} = 40.58$
For $x^2 = 40.58$ and $df = 4$, p < .0001. Since .0001 < .05, we will reject the null hypothesis. Thus, the percentage of students using each bank in 2009 is different from the percentage of students using each bank in 2008.

26. B: Since there are two factors at play: curricula and placement with teacher, the outcome can be influenced by either the treatment (curricula) or other blocks associated

with Teacher A and Teacher B. This randomized block design could be broken up according to teacher (Teacher A or Teacher B) and then further broken into blocks for each of the three curricula per teacher.
Choice A is not correct because the problem does state that the students are broken up according to teacher. With a completely randomized design, patients are randomly assigned to each group. This process differs from randomized block design, since the latter will randomly assign students taught by each teacher to the three groups of curricula.
Choice C is not correct because there is not a pre-post situation occurring. You are not examining a pre-test, the effect of the curriculum, and a post-test.
Choice D is not correct because we are not giving the same curricula time and time again, or giving it to different groups than the ones described. Replication is used to determine if you will get similar results the second time around.

27. There are limitless examples of the Law of Large Numbers. One such example would be the number the arrow on a spinner points to after each spin. If the spinner is divided into 8 sections, with 8 numbers, labeled 1 - 8, there is a 1/8 chance of the arrow landing on a 3. If we spin the spinner 10 times, we will get an experimental probability that approaches the theoretical probability of 1/8. If we spin the spinner 100 times, the experimental probability will be even closer to 1/8. Thus, the law states that as the repetitions of an experiment increase, the experimental probability mirrors or approaches the theoretical probability.

28. C: In order to calculate the correlation coefficient, we can either use the calculator or compute manually.
Using the calculator, you simply enter the x-values into L1 and enter the y-values into L2. Then, go to Stat, Calculate, and lastly LinReg (a + bx). Enter L1, L2 and click "enter". Doing so will give you the r, or correlation coefficient. The correlation coefficient is .68.
You can also calculate the correlation coefficient manually. First, calculate the sum of squares for x and sum of squares for y (68,412 and 69,985, respectively. Then find $\bar{X}$ and $\bar{Y}$.
The sample mean of x ($\bar{X}$) can be found by summing the x-values and dividing by n:
$\frac{824}{10} = 82.4$
The sample mean of y ($\bar{Y}$) can be found by summing the y-values and dividing by n:
$\frac{833}{10} = 83.3$
The formula for the correlation coefficient is:
$$\hat{p} = \frac{S_{xy}}{\sqrt{S_{xx}S_{yy}}}$$

So, we need to find each part.
S_{xx} is found by starting with the raw sum of squares for x and subtracting the quantity (n multiplied by ($\bar{X}^2$)). In other words, you have:
$S_{xx} = 68412 - 10(6789.76) = 514.4$
S_{yy} is found by starting with the raw sum of squares for y and subtracting the quantity (n multiplied by ($\bar{Y}^2$)). In other words, you have:
$S_{yy} = 69985 - 10(6938.89) = 596.1$
S_{xy} is found by summing the products of $(X - \bar{X})(Y - \bar{Y})$. In other words, you have:
$S_{xy} = 374.8$
Now, we can substitute our values back into the correlation coefficient formula:
$$\hat{p} = \frac{S_{xy}}{\sqrt{S_{xx}S_{yy}}} = \frac{374.8}{\sqrt{(514.4)(596.1)}} \approx .68$$

Again, we found our correlation coefficient to be .68.

29. D: The formula for finding the nth percentile is:

$$l = \frac{(n+1)k}{100}$$

Substituting in our values, we have:

$l = \frac{(21)85}{100} = 17.85$

So, the value at the 85th percentile is between the 17th and 18th recorded mileage. We must list the data in order from least to greatest.

The 17th recorded mileage is 51, and the 18th recorded mileage is 56. Adjusting for our decimal, we find the 17.85th value to be 55.25. (The range in values is 5, and .85 × 5 = 4.25; 4.25 added to 51 is 55.25).

Thus, 55 miles or less are driven on 85% of the days; 56 miles or more are driven on the other 15% of the days.

30. C: In order to determine the expected number of graduate students with a drama preference, you must calculate the association between the two categories.

Thus, $\frac{(142)(77)}{299} \approx 36.57$. Notice, we took the row total associated with the graduate category, multiplied it by the column total for drama preference, and divided the product by the total sample size.

31. C: We must first find the standard deviation for the sample mean.

$\sigma_{\bar{X}} = \frac{.75}{\sqrt{65}} = .09$

Next, we will find the area or probability between the two values.

$P(2.85 < \bar{X} < 3.25)$

$=P(\frac{2.85-2.99)}{.09} < Z < \frac{3.25-2.99}{.09}$

$=P(-1.56 < Z < 2.89)$

$\approx .95$

32. D: The point estimate for the population variance is the sample variance. The formula for computing sample variance is as follows:

$s^2 = \frac{\sum_{i=1}^{n}(X_i - \bar{X})^2}{n-1}$

The sample mean is approximately 695.17 and the numerator is calculated to be 8,009,909.67. The denominator is 11 (12 – 1). Thus,

$s^2 = \frac{8009909.67}{11} \approx 728{,}173.6$

33. B: We know we can use a z-interval because the samples are independent and randomly selected; also, $n\hat{p} > 10$ and $n(1-\hat{p}) > 10$.

We must first find the sample proportion, which can be found by dividing the number of defective cars by the total number of cars sampled. So,

$\hat{p} = \frac{26}{150} \approx .17$

We can construct the 95% confidence interval by computing:

$Z_{\alpha/2}\sqrt{\frac{\hat{p}(1-\hat{p})}{n}}$

Note that $Z_{.025} = 1.96$.

So, we have

$$Z_{\alpha/2}\sqrt{\frac{\hat{p}(1-\hat{p})}{n}} = 1.96\sqrt{\frac{.17(1-.17)}{150}} = .06$$

$$.17 \pm .06 = .11, .23$$

Thus, the true proportion is between .11 and .23.
You can also compute the confidence interval on a graphing calculator.
Go to STAT
Scroll over to Tests and choose 1-PropZInt
Enter $x = 26$ and $n = 150$
Choose Confidence level of 95%.
Choose Calculate
This displays a 95% confidence interval of .11 to .23.

34. A: We must first find the sample proportion for each, which can be found by dividing the number of students who intend to go to the university by the total number of students sampled. So, the sample proportion for Wichita Falls is:

$$\hat{p}_1 = \frac{45}{80} \approx .56$$

The sample proportion for Corpus Christi is:

$$\hat{p}_2 = \frac{39}{75} \approx .52$$

We can construct the 90% confidence interval by computing:

$$Z_{\alpha/2}\sqrt{\frac{\hat{p}_1(1-\hat{p}_1)}{n_1} + \frac{\hat{p}_2(1-\hat{p}_2)}{n_2}}$$

Note that $Z_{.05} = 1.645$.
So, we have

$$Z_{\alpha/2}\sqrt{\frac{\hat{p}_1(1-\hat{p}_1)}{n_1} + \frac{\hat{p}_2(1-\hat{p}_2)}{n_2}} = 1.645\sqrt{\frac{.56(1-.56)}{80} + \frac{.52(1-.52)}{75}} \approx .13$$

We must add and subtract this amount to and from the difference between the proportions.
.56 – .52 = .04
So,

$$.04 \pm .13 = -.09, .17$$

Thus, the difference between the proportions is between –.09 and .17.
You can also compute the confidence interval on a graphing calculator.
Go to STAT
Scroll over to Tests and choose 2-PropZInt
Enter $x1 = 45$ and $n1 = 80; x2 = 39$ and $n2 = 75$
Choose Confidence level of 90%.
Choose Calculate
This displays a 90% confidence interval of –.09 to .17.

35. C: We can assume the sample is representative of all of the doctor's insomnia patients. A histogram of the differences between the number of hours slept during the weeks of receiving the treatment and the weeks of receiving the placebo appeared to be normally distributed.

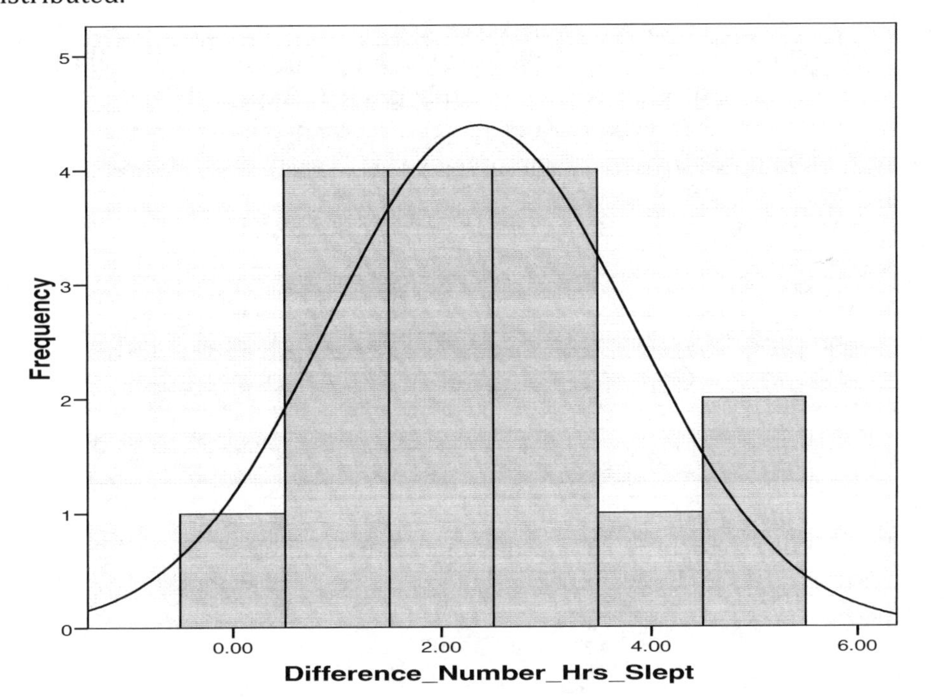

We need to find the sample mean of the differences. Thus,
$\bar{D} = 2.375$
We also need to find the standard deviation of the differences.
$S_D \approx 1.45$
Note. Our degrees of freedom equals 15.
The critical t-value for 15 degrees of freedom, with .025 in each tail, is 2.131.
$t_{.025}(15)\left(\frac{S_D}{\sqrt{n}}\right) = 2.131\left(\frac{1.45}{\sqrt{16}}\right) \approx .77$ hours of sleep per night
Thus, if we add and subtract this value to and from our sample mean of the differences, we get:
$2.375 \pm .77 = 1.61, 3.15$
So, for 95% confidence, we can say the true mean difference in hours slept is between 1.61 and 3.15 hours per night.
We can also compute the t-interval using a graphing calculator.
Enter the values into the L1 and L2 lists
Create a difference column, L3
Go to STAT
Scroll over to Tests and choose TInt
Enter L3
Choose Confidence level of 95%.
Choose Calculate
We get a t-interval of 1.6 to 3.15, which is slightly difference from our manual calculation, due to rounding.

36. With 40 students selected per school and 30 schools examined, there are 1200 students involved.

The number of students not passing can be found with the following calculation:
$(6(0) + 8(1) + 4(2) + 3(3) + 9(4)) = 61$
The probability of students not passing, or $\hat{p}$, can be found by dividing the number of students not passing by the total number of students.
Thus, $61/1200 \approx .05$.
Our null hypothesis states that a binomial distribution is an appropriate model, while the alternative hypothesis states that a binomial distribution is not an appropriate model.
A chi-squared test can be used to determine which is correct. We need to compute the probability that each number of students not passing the test. We can do so by using the following formula:
$P(X) = \binom{n}{r}\hat{p}^r(1-\hat{p})^{n-r}$

# of students not passing	0	1	2	3	4	5+
Probability (X)	.13	.27	.28	.19	.09	.04
Expected # of students	3.9	8.1	8.4	5.7	2.7	1.2

We will combine the last three columns, so the chi-squared test will work properly.

# of students not passing	0	1	2	3+
Observed # of students	6	8	4	12
Expected # of students	3.9	8.1	8.4	9.6

For degrees of freedom of 3 and level of significance of .05, the critical x^2 value is 7.81.
Computing the x^2 value gives:
$x^2 = \sum_{i=1}^{4} \frac{(O_i - E_i)^2}{E_i}$
$= \frac{(6-3.9)^2}{3.9} + \frac{(8-8.1)^2}{8.1} + \frac{(4-8.4)^2}{8.4} + \frac{(12-9.6)^2}{9.6}$
$= 4.03$
Since the observed chi-squared value is less than the critical chi-squared value, we will fail to reject the null hypothesis. In other words, a binomial distribution is a good model for the data.

37. With a matched pairs design, participant results are matched per person, or per appropriate pair. In other words, the results from two treatments for matched persons, or the results found from a pre-test and post-test for the same person are used in the analysis. (You might pair a husband and wife, participants of the same age, etc.) One such scenario whereby a matched pairs design would be used is when a teacher wishes to assess the effects of lesson inclusion of algebra tiles on student understanding of factoring expressions. Each student's understanding prior to the intervention would be measured and then compared to the same student's understanding after the intervention.

38. B: In order to calculate the interquartile range, you must first calculate Q1 and Q3. Q1 can be found by determining the median of the lower half of the observations. Q1 = the average of 140 and 150.
$\frac{140+150}{2} = 145$

Q3 can be found by determining the median of the upper half of the observations. Q3 = the average of 240 and 290.

$\frac{240+290}{2} = 265$

The interquartile range equals Q3 – Q1, or 265 – 145, in this case, which equals 120.
You can also determine the quartiles by entering the data into a list on the graphing calculator and finding the 1-Var statistics under the STAT button.

39 A: In order to find the number of writers that spend less than $70 per month on coffee, you simply find the 70 on the *x*-axis and go over to the corresponding value on the *y*-axis. The closest cumulative frequency approximation for $70 is 33 writers.

40. We would reject the null hypothesis, thus concluding there is a significant difference between the two means.
In order to solve the problem, we must calculate the standard error and observed t-statistic.

$SE = \frac{s}{\sqrt{n}} = \frac{2.5}{\sqrt{35}} \approx .42$

$t = \left(\frac{\bar{X}-\mu}{SE}\right) = \left(\frac{102-97}{.42}\right) \approx 11.9$

The observed t-value is approximately 11.9. The critical t-value for 34 degrees of freedom is around 1.684. Since the observed t-value is greater than the critical t-value, we will reject the null hypothesis, thus concluding a difference between the sample mean and population mean.

41. D: We must first calculate the pooled sample proportion.

$p = \frac{(p_1 n_1 + p_2 n_2)}{n_1 + n_2}$

$= \frac{(.04)(1216)+(.06)(1310)}{2526} \approx .05$

Next, we will calculate the standard error.

$SE = \sqrt{p(1-p)\frac{n_1+n_2}{n_1 n_2}}$

$= \sqrt{.05(1-.05)(.002)} \approx .0086$

$Z = \frac{p_1 - p_2}{SE} = \frac{-.02}{.0086} \approx -2.3$

The z-value results in an area less than .05. Thus, we will reject the null hypothesis and conclude a significant difference in proportions.

42. We must first determine the least square regression equation, in order to compute the predicted GMAT scores.
We can enter the x-values into L1 and enter the y-values into L2 on the graphing calculator.
Next, we do the following:
Go to STAT
Choose Calculate
Choose LinReg (a+bx)
Enter L1, L2
Press Enter
We get the following output:
$a \approx 161.37$
$b \approx .27$
Thus, the least squares regression equation is $y = 161.37 + .27x$.
We can now use our least squares regression equation to find the predicted GMAT scores.

Let's create a table to help with the process.

SAT Scores	GMAT Scores	Predicted GMAT Scores $\hat{y} = 161.37 + .27x$	Residual $e = y - \hat{y}$
820	450	382.77	67.23
1260	600	501.57	98.43
2100	780	728.37	51.63
960	480	420.57	59.43
680	260	344.97	−84.97
1450	620	552.87	67.13
1310	580	515.07	64.93
1020	360	436.77	−76.77
1680	410	614.97	−204.97
2060	710	717.57	−7.57
1810	690	650.07	39.93
1430	440	547.47	−107.47

The residual plot should look like:

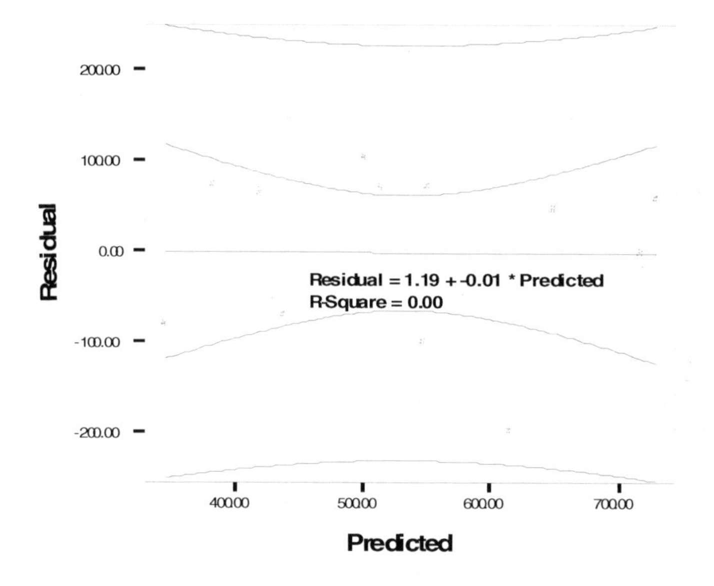

43. C: A reciprocal transformation is used when rather large values of *x* are included in the analysis, in order to minimize the effect of those large numbers.
Choices A and D are not correct because they deal with increasing and decreasing slopes.
Choice B is not correct because we are not given any information about spread, as related to increases with the mean.

44. The back-to-back stemplot looks like:

Region A		**Region B**
Leaves	**Stems**	**Leaves**
7	1	0, 9, 1, 6, 8
2, 9	2	9, 3, 3, 6, 9
4	3	6, 5, 5, 4, 0, 1
3, 1	4	3, 8, 0, 2, 2
6, 1, 0, 3	5	9, 1, 6, 2
2, 7, 8, 4, 9	6	4, 8
5, 3, 1, 2, 2	7	3
6, 2, 9, 3, 6	8	6, 1
0, 2, 5, 1, 4	9	

The distribution for number of miles traveled by the Region A sales personnel is skewed left. The distribution for number of miles traveled by the Region B sales personnel is skewed right. In other words, the Region A sales personnel travel more than the personnel in Region B.

45. B: We need to find the sample mean. Thus,
$\bar{X} \approx 34.04$
We also need to find the standard deviation.
$s \approx 1.13$
Note. Our degrees of freedom equals 29.
The critical t-value for 29 degrees of freedom, with .05 in each tail, is 1.699.
$t_{.05}(29)\left(\frac{s}{\sqrt{n}}\right) = 1.699\left(\frac{1.13}{\sqrt{30}}\right) \approx .35$
Thus, if we add and subtract this value to and from our sample mean, we get:
$34.04 \pm .35 = 33.69, 34.39$
So, for 90% confidence, we can say the true mean gas mileage is between 33.69 and 34.39 miles per gallon.
We can also compute the t-interval using a graphing calculator.
Enter the values into the L1 list
Go to STAT
Scroll over to Tests and choose TInt
Enter L1
Choose Confidence level of 90%.
Choose Calculate

46. In order to determine whether or not there is homogeneity across the proportions, we must use a chi-square test of homogeneity of proportions.
First, we will state our null and alternative hypotheses:
H_0: All proportions are equal; the proportion of employees who answered "yes" is the same across all employee levels or categories.
H_a: At least two employee categories differ in the proportion of those who answered "yes".
We will need to determine the total number of employees who answered "yes" and "no". So, we will add another row at the bottom of our table. We will rename the "n column" as "Row Total" and label the sample size.

	Yes	No	Row Total
Intern	24	11	35
Entry-Level	29	6	35
Mid-Level	38	7	45
Senior Level	31	14	45
Column Total	122	38	n = 160

We will need to determine the expected number of responses for each employee category.

$E = \frac{R_i C_j}{n}$

We will create a new table with the expected number of responses below:

	Expected "Yes" Counts	Expected "No" Counts
Intern	26.69	8.31
Entry-Level	26.69	8.31
Mid-Level	34.31	10.69
Senior Level	34.31	10.69

Since all counts are greater than 5, we can use a chi-squared test. Our degrees of freedom equal 3, since there are four categories.

$x^2 = \sum \frac{(O_{r,c} - E_{r,c})^2}{E_{r,c}} \approx 5$

The critical x^2-value for 3 df and .05 level of significance is 7.81.

Since our observed x^2-value is less than 7.81, we will fail to reject the null hypothesis, thus concluding that there is homogeneity of proportions. In other words, the responses do not vary across the proportions of employees interviewed.

47. C: In order for a t-test of the slope of a least squares regression line to be performed, the standard deviation of residuals should be constant. The condition does not state that the standard deviation of the residuals be 1. The other three choices are required conditions for the test.

48. B: We must determine the difference between the proportions, using pooled standard deviation and t-statistic.

First of all, we should note that we have 91 degrees of freedom (45 + 48 – 2); also, since we're using .05 as the significance level, our critical t-value for 91 degrees of freedom, is approximately 1.66.

Now, we need to calculate the pooled standard deviation.

$s_p = \sqrt{\frac{(45-1)43.8+(48-1)45.9}{45+48-2}}$

$= \sqrt{\frac{4084.5}{91}}$

≈ 6.7

Next, we need to calculate the observed t-value.

$t = \frac{(\bar{X}_1 - \bar{X}_2) - 0}{s_p \sqrt{\left(\frac{1}{n_1} + \frac{1}{n_2}\right)}}$

$= \frac{(87.2 - 85.6) - 0}{6.7 \sqrt{\frac{1}{45} + \frac{1}{48}}}$

≈ 1.15

The observed t-value is approximately 1.15 and is less than the critical t-value of 1.66, thus we will fail to reject the null hypothesis, which states there is no difference between the two means. Thus, there is no significant difference between the two groups, due to the intervention described.

49. You will fail to reject the null hypothesis, or the hypothesis that states that Robert sells 75% of the vacuum cleaners.

In order to make this determination, you must first identify the two proportions: p and q.

p = .75 and q = .25

The sample proportions are:

$\hat{p} = \frac{45}{70} \approx .64$ and

$\hat{q} = \frac{25}{70} \approx .36$

The standard deviation can be found by using the following formula:

$\sigma = \sqrt{\frac{p \cdot q}{n}} = \sqrt{\frac{.75 \cdot .25}{70}} \approx .05$

$Z = \frac{\hat{p} - p}{\sigma} = \frac{.64 - .75}{.05} = -2.2$

The area or probability is approximately .01, which is less than .05, so we will reject the null hypothesis and declare significance. In other words, we will reject the claim that he sells 75% of the vacuum cleaners.

50. A: You must first calculate $\hat{p}$ and $\hat{q}$.

$\hat{p} = \frac{321}{450} \approx .71$

$\hat{q} = 1 - \hat{p} = 1 - .71 = .29$

Now, we can calculate the standard deviation of the sample proportions:

$\sigma = \sqrt{\left(\hat{p} \cdot \frac{\hat{q}}{n}\right)}$

$= \sqrt{\left(.71 \cdot \frac{.29}{450}\right)}$

$\approx .02$

Secret Key #1 - Time is Your Greatest Enemy

Pace Yourself

Wear a watch. At the beginning of the test, check the time (or start a chronometer on your watch to count the minutes), and check the time after every few questions to make sure you are "on schedule."

If you are forced to speed up, do it efficiently. Usually one or more answer choices can be eliminated without too much difficulty. Above all, don't panic. Don't speed up and just begin guessing at random choices. By pacing yourself, and continually monitoring your progress against your watch, you will always know exactly how far ahead or behind you are with your available time. If you find that you are one minute behind on the test, don't skip one question without spending any time on it, just to catch back up. Take 15 fewer seconds on the next four questions, and after four questions you'll have caught back up. Once you catch back up, you can continue working each problem at your normal pace.

Furthermore, don't dwell on the problems that you were rushed on. If a problem was taking up too much time and you made a hurried guess, it must be difficult. The difficult questions are the ones you are most likely to miss anyway, so it isn't a big loss. It is better to end with more time than you need than to run out of time.

Lastly, sometimes it is beneficial to slow down if you are constantly getting ahead of time. You are always more likely to catch a careless mistake by working more slowly than quickly, and among very high-scoring test takers (those who are likely to have lots of time left over), careless errors affect the score more than mastery of material.

Secret Key #2 - Guessing is not Guesswork

You probably know that guessing is a good idea. Unlike other standardized tests, there is no penalty for getting a wrong answer. Even if you have no idea about a question, you still have a 20-25% chance of getting it right.

Most test takers do not understand the impact that proper guessing can have on their score. Unless you score extremely high, guessing will significantly contribute to your final score.

Monkeys Take the Test

What most test takers don't realize is that to insure that 20-25% chance, you have to guess randomly. If you put 20 monkeys in a room to take this test, assuming they answered once per question and behaved themselves, on average they would get 20-25% of the questions correct. Put 20 test takers in the room, and the average will be much lower among guessed questions. Why?

1. The test writers intentionally write deceptive answer choices that "look" right. A test

taker has no idea about a question, so he picks the "best looking" answer, which is often wrong. The monkey has no idea what looks good and what doesn't, so it will consistently be right about 20-25% of the time.
2. Test takers will eliminate answer choices from the guessing pool based on a hunch or intuition. Simple but correct answers often get excluded, leaving a 0% chance of being correct. The monkey has no clue, and often gets lucky with the best choice.

This is why the process of elimination endorsed by most test courses is flawed and detrimental to your performance. Test takers don't guess; they make an ignorant stab in the dark that is usually worse than random.

$5 Challenge

Let me introduce one of the most valuable ideas of this course—the $5 challenge:

You only mark your "best guess" if you are willing to bet $5 on it.
You only eliminate choices from guessing if you are willing to bet $5 on it.

Why $5? Five dollars is an amount of money that is small yet not insignificant, and can really add up fast (20 questions could cost you $100). Likewise, each answer choice on one question of the test will have a small impact on your overall score, but it can really add up to a lot of points in the end.

The process of elimination IS valuable. The following shows your chance of guessing it right:

If you eliminate wrong answer choices until only this many remain:	Chance of getting it correct:
1	100%
2	50%
3	33%

However, if you accidentally eliminate the right answer or go on a hunch for an incorrect answer, your chances drop dramatically—to 0%. By guessing among all the answer choices, you are GUARANTEED to have a shot at the right answer.

That's why the $5 test is so valuable. If you give up the advantage and safety of a pure guess, it had better be worth the risk.

What we still haven't covered is how to be sure that whatever guess you make is truly random. Here's the easiest way:

Always pick the first answer choice among those remaining

Such a technique means that you have decided, **before you see a single test question**, exactly how you are going to guess, and since the order of choices tells you nothing about which one is correct, this guessing technique is perfectly random.

This section is not meant to scare you away from making educated guesses or eliminating choices; you just need to define when a choice is worth eliminating. The $5 test, along with a pre-defined random guessing strategy, is the best way to make sure you reap all of the benefits of guessing.

Secret Key #3 - Practice Smarter, Not Harder

Many test takers delay the test preparation process because they dread the awful amounts of practice time they think necessary to succeed on the test. We have refined an effective method that will take you only a fraction of the time.

There are a number of "obstacles" in the path to success. Among these are answering questions, finishing in time, and mastering test-taking strategies. All must be executed on the day of the test at peak performance, or your score will suffer. The test is a mental marathon that has a large impact on your future.
Just like a marathon runner, it is important to work your way up to the full challenge. So first you just worry about questions, and then time, and finally strategy:

Success Strategy

1. Find a good source for practice tests.
2. If you are willing to make a larger time investment, consider using more than one study guide. Often the different approaches of multiple authors will help you "get" difficult concepts.
3. Take a practice test with no time constraints, with all study helps, "open book." Take your time with questions and focus on applying strategies.
4. Take a practice test with time constraints, with all guides, "open book."
5. Take a final practice test without open material and with time limits.

If you have time to take more practice tests, just repeat step 5. By gradually exposing yourself to the full rigors of the test environment, you will condition your mind to the stress of test day and maximize your success.

Secret Key #4 - Prepare, Don't Procrastinate

Let me state an obvious fact: if you take the test three times, you will probably get three different scores. This is due to the way you feel on test day, the level of preparedness you have, and the version of the test you see. Despite the test writers' claims to the contrary, some versions of the test WILL be easier for you than others.
Since your future depends so much on your score, you should maximize your chances of success. In order to maximize the likelihood of success, you've got to prepare in advance. This means taking practice tests and spending time learning the information and test taking

strategies you will need to succeed.

Never go take the actual test as a "practice" test, expecting that you can just take it again if you need to. Take all the practice tests you can on your own, but when you go to take the official test, be prepared, be focused, and do your best the first time!

Secret Key #5 - Test Yourself

Everyone knows that time is money. There is no need to spend too much of your time or too little of your time preparing for the test. You should only spend as much of your precious time preparing as is necessary for you to get the score you need.

Once you have taken a practice test under real conditions of time constraints, then you will know if you are ready for the test or not.

If you have scored extremely high the first time that you take the practice test, then there is not much point in spending countless hours studying. You are already there.

Benchmark your abilities by retaking practice tests and seeing how much you have improved. Once you consistently score high enough to guarantee success, then you are ready.

If you have scored well below where you need, then knuckle down and begin studying in earnest. Check your improvement regularly through the use of practice tests under real conditions. Above all, don't worry, panic, or give up. The key is perseverance!

Then, when you go to take the test, remain confident and remember how well you did on the practice tests. If you can score high enough on a practice test, then you can do the same on the real thing.

General Strategies

The most important thing you can do is to ignore your fears and jump into the test immediately. Do not be overwhelmed by any strange-sounding terms. You have to jump into the test like jumping into a pool—all at once is the easiest way.

Make Predictions
As you read and understand the question, try to guess what the answer will be. Remember that several of the answer choices are wrong, and once you begin reading them, your mind will immediately become cluttered with answer choices designed to throw you off. Your mind is typically the most focused immediately after you have read the question and digested its contents. If you can, try to predict what the correct answer will be. You may be surprised at what you can predict.

Quickly scan the choices and see if your prediction is in the listed answer choices. If it is, then you can be quite confident that you have the right answer. It still won't hurt to check the other answer choices, but most of the time, you've got it!

Answer the Question

It may seem obvious to only pick answer choices that answer the question, but the test writers can create some excellent answer choices that are wrong. Don't pick an answer just because it sounds right, or you believe it to be true. It MUST answer the question. Once you've made your selection, always go back and check it against the question and make sure that you didn't misread the question and that the answer choice does answer the question posed.

Benchmark

After you read the first answer choice, decide if you think it sounds correct or not. If it doesn't, move on to the next answer choice. If it does, mentally mark that answer choice. This doesn't mean that you've definitely selected it as your answer choice, it just means that it's the best you've seen thus far. Go ahead and read the next choice. If the next choice is worse than the one you've already selected, keep going to the next answer choice. If the next choice is better than the choice you've already selected, mentally mark the new answer choice as your best guess.

The first answer choice that you select becomes your standard. Every other answer choice must be benchmarked against that standard. That choice is correct until proven otherwise by another answer choice beating it out. Once you've decided that no other answer choice seems as good, do one final check to ensure that your answer choice answers the question posed.

Valid Information

Don't discount any of the information provided in the question. Every piece of information may be necessary to determine the correct answer. None of the information in the question is there to throw you off (while the answer choices will certainly have information to throw you off). If two seemingly unrelated topics are discussed, don't ignore either. You can be confident there is a relationship, or it wouldn't be included in the question, and you are probably going to have to determine what is that relationship to find the answer.

Avoid "Fact Traps"

Don't get distracted by a choice that is factually true. Your search is for the answer that answers the question. Stay focused and don't fall for an answer that is true but irrelevant. Always go back to the question and make sure you're choosing an answer that actually answers the question and is not just a true statement. An answer can be factually correct, but it MUST answer the question asked. Additionally, two answers can both be seemingly correct, so be sure to read all of the answer choices, and make sure that you get the one that BEST answers the question.

Milk the Question

Some of the questions may throw you completely off. They might deal with a subject you have not been exposed to, or one that you haven't reviewed in years. While your lack of knowledge about the subject will be a hindrance, the question itself can give you many clues that will help you find the correct answer. Read the question carefully and look for clues. Watch particularly for adjectives and nouns describing difficult terms or words that you don't recognize. Regardless of whether you completely understand a word or not, replacing it with a synonym, either provided or one you more familiar with, may help you to understand what the questions are asking. Rather than wracking your mind about specific detailed information concerning a difficult term or word, try to use mental substitutes that are easier to understand.

The Trap of Familiarity
Don't just choose a word because you recognize it. On difficult questions, you may not recognize a number of words in the answer choices. The test writers don't put "make-believe" words on the test, so don't think that just because you only recognize all the words in one answer choice that that answer choice must be correct. If you only recognize words in one answer choice, then focus on that one. Is it correct? Try your best to determine if it is correct. If it is, that's great. If not, eliminate it. Each word and answer choice you eliminate increases your chances of getting the question correct, even if you then have to guess among the unfamiliar choices.

Eliminate Answers
Eliminate choices as soon as you realize they are wrong. But be careful! Make sure you consider all of the possible answer choices. Just because one appears right, doesn't mean that the next one won't be even better! The test writers will usually put more than one good answer choice for every question, so read all of them. Don't worry if you are stuck between two that seem right. By getting down to just two remaining possible choices, your odds are now 50/50. Rather than wasting too much time, play the odds. You are guessing, but guessing wisely because you've been able to knock out some of the answer choices that you know are wrong. If you are eliminating choices and realize that the last answer choice you are left with is also obviously wrong, don't panic. Start over and consider each choice again. There may easily be something that you missed the first time and will realize on the second pass.

Tough Questions
If you are stumped on a problem or it appears too hard or too difficult, don't waste time. Move on! Remember though, if you can quickly check for obviously incorrect answer choices, your chances of guessing correctly are greatly improved. Before you completely give up, at least try to knock out a couple of possible answers. Eliminate what you can and then guess at the remaining answer choices before moving on.

Brainstorm
If you get stuck on a difficult question, spend a few seconds quickly brainstorming. Run through the complete list of possible answer choices. Look at each choice and ask yourself, "Could this answer the question satisfactorily?" Go through each answer choice and consider it independently of the others. By systematically going through all possibilities, you may find something that you would otherwise overlook. Remember though that when you get stuck, it's important to try to keep moving.

Read Carefully
Understand the problem. Read the question and answer choices carefully. Don't miss the question because you misread the terms. You have plenty of time to read each question thoroughly and make sure you understand what is being asked. Yet a happy medium must be attained, so don't waste too much time. You must read carefully, but efficiently.

Face Value
When in doubt, use common sense. Always accept the situation in the problem at face value. Don't read too much into it. These problems will not require you to make huge leaps

of logic. The test writers aren't trying to throw you off with a cheap trick. If you have to go beyond creativity and make a leap of logic in order to have an answer choice answer the question, then you should look at the other answer choices. Don't overcomplicate the problem by creating theoretical relationships or explanations that will warp time or space. These are normal problems rooted in reality. It's just that the applicable relationship or explanation may not be readily apparent and you have to figure things out. Use your common sense to interpret anything that isn't clear.

Prefixes

If you're having trouble with a word in the question or answer choices, try dissecting it. Take advantage of every clue that the word might include. Prefixes and suffixes can be a huge help. Usually they allow you to determine a basic meaning. Pre- means before, post- means after, pro - is positive, de- is negative. From these prefixes and suffixes, you can get an idea of the general meaning of the word and try to put it into context. Beware though of any traps. Just because con- is the opposite of pro-, doesn't necessarily mean congress is the opposite of progress!

Hedge Phrases

Watch out for critical hedge phrases, led off with words such as "likely," "may," "can," "sometimes," "often," "almost," "mostly," "usually," "generally," "rarely," and "sometimes." Question writers insert these hedge phrases to cover every possibility. Often an answer choice will be wrong simply because it leaves no room for exception. Unless the situation calls for them, avoid answer choices that have definitive words like "exactly," and "always."

Switchback Words

Stay alert for "switchbacks." These are the words and phrases frequently used to alert you to shifts in thought. The most common switchback word is "but." Others include "although," "however," "nevertheless," "on the other hand," "even though," "while," "in spite of," "despite," and "regardless of."

New Information

Correct answer choices will rarely have completely new information included. Answer choices typically are straightforward reflections of the material asked about and will directly relate to the question. If a new piece of information is included in an answer choice that doesn't even seem to relate to the topic being asked about, then that answer choice is likely incorrect. All of the information needed to answer the question is usually provided for you in the question. You should not have to make guesses that are unsupported or choose answer choices that require unknown information that cannot be reasoned from what is given.

Time Management

On technical questions, don't get lost on the technical terms. Don't spend too much time on any one question. If you don't know what a term means, then odds are you aren't going to get much further since you don't have a dictionary. You should be able to immediately recognize whether or not you know a term. If you don't, work with the other clues that you have—the other answer choices and terms provided—but don't waste too much time trying to figure out a difficult term that you don't know.

Contextual Clues

Look for contextual clues. An answer can be right but not the correct answer. The contextual clues will help you find the answer that is most right and is correct. Understand the context in which a phrase or statement is made. This will help you make important distinctions.

Don't Panic

Panicking will not answer any questions for you; therefore, it isn't helpful. When you first see the question, if your mind goes blank, take a deep breath. Force yourself to mechanically go through the steps of solving the problem using the strategies you've learned.

Pace Yourself

Don't get clock fever. It's easy to be overwhelmed when you're looking at a page full of questions, your mind is full of random thoughts and feeling confused, and the clock is ticking down faster than you would like. Calm down and maintain the pace that you have set for yourself. As long as you are on track by monitoring your pace, you are guaranteed to have enough time for yourself. When you get to the last few minutes of the test, it may seem like you won't have enough time left, but if you only have as many questions as you should have left at that point, then you're right on track!

Answer Selection

The best way to pick an answer choice is to eliminate all of those that are wrong, until only one is left and confirm that is the correct answer. Sometimes though, an answer choice may immediately look right. Be careful! Take a second to make sure that the other choices are not equally obvious. Don't make a hasty mistake. There are only two times that you should stop before checking other answers. First is when you are positive that the answer choice you have selected is correct. Second is when time is almost out and you have to make a quick guess!

Check Your Work

Since you will probably not know every term listed and the answer to every question, it is important that you get credit for the ones that you do know. Don't miss any questions through careless mistakes. If at all possible, try to take a second to look back over your answer selection and make sure you've selected the correct answer choice and haven't made a costly careless mistake (such as marking an answer choice that you didn't mean to mark). The time it takes for this quick double check should more than pay for itself in caught mistakes.

Beware of Directly Quoted Answers

Sometimes an answer choice will repeat word for word a portion of the question or reference section. However, beware of such exact duplication. It may be a trap! More than likely, the correct choice will paraphrase or summarize a point, rather than being exactly the same wording.

Slang

Scientific sounding answers are better than slang ones. An answer choice that begins "To compare the outcomes..." is much more likely to be correct than one that begins "Because some people insisted..."

Extreme Statements

Avoid wild answers that throw out highly controversial ideas that are proclaimed as established fact. An answer choice that states the "process should be used in certain situations, if..." is much more likely to be correct than one that states the "process should be discontinued completely." The first is a calm rational statement and doesn't even make a definitive, uncompromising stance, using a hedge word "if" to provide wiggle room, whereas the second choice is a radical idea and far more extreme.

Answer Choice Families

When you have two or more answer choices that are direct opposites or parallels, one of them is usually the correct answer. For instance, if one answer choice states "x increases" and another answer choice states "x decreases" or "y increases," then those two or three answer choices are very similar in construction and fall into the same family of answer choices. A family of answer choices consists of two or three answer choices, very similar in construction, but often with directly opposite meanings. Usually the correct answer choice will be in that family of answer choices. The "odd man out" or answer choice that doesn't seem to fit the parallel construction of the other answer choices is more likely to be incorrect.

Special Report: What Your Test Score Will Tell You About Your IQ

Did you know that most standardized tests correlate very strongly with IQ? In fact, your general intelligence is a better predictor of your success than any other factor, and most tests intentionally measure this trait to some degree to ensure that those selected by the test are truly qualified for the test's purposes.

Before we can delve into the relation between your test score and IQ, I will first have to explain what exactly is IQ. Here's the formula:

Your IQ = 100 + (Number of standard deviations below or above the average)*15

Now, let's define standard deviations by using an example. If we have 5 people with 5 different heights, then first we calculate the average. Let's say the average was 65 inches. The standard deviation is the "average distance" away from the average of each of the members. It is a direct measure of variability. If the 5 people included Jackie Chan and Shaquille O'Neal, obviously there's a lot more variability in that group than a group of 5 sisters who are all within 6 inches in height of each other. The standard deviation uses a number to characterize the average range of difference within a group.

A convenient feature of most groups is that they have a "normal" distribution. It makes sense that most things would be normal, right? Without getting into a bunch of statistical mumbo-jumbo, you just need to know that if you know the average of the group and the standard deviation, you can successfully predict someone's percentile rank in the group.

Confused? Let me give you an example. If instead of 5 people's heights, we had 100 people, we could figure out their rank in height JUST by knowing the average, standard deviation, and their height. We wouldn't need to know each person's height and manually rank them, we could just predict their rank based on three numbers.

What this means is that you can take your PERCENTILE rank that is often given with your test and relate this to your RELATIVE IQ of people taking the test - that is, your IQ relative to the people taking the test. Obviously, there's no way to know your actual IQ because the people taking a standardized test are usually not very good samples of the general population. Many of those with extremely low IQ's never achieve a level of success or competency necessary to complete a typical standardized test. In fact, professional psychologists who measure IQ actually have to use non-written tests that can fairly measure the IQ of those not able to complete a traditional test.

The bottom line is to not take your test score too seriously, but it is fun to compute your "relative IQ" among the people who took the test with you. I've done the calculations below. Just look up your percentile rank in the left and then you'll see your "relative IQ" for your test in the right hand column.

Percentile Rank	Your Relative IQ		Percentile Rank	Your Relative IQ
99	135		59	103
98	131		58	103
97	128		57	103
96	126		56	102
95	125		55	102
94	123		54	102
93	122		53	101
92	121		52	101
91	120		51	100
90	119		50	100
89	118		49	100
88	118		48	99
87	117		47	99
86	116		46	98
85	116		45	98
84	115		44	98
83	114		43	97
82	114		42	97
81	113		41	97
80	113		40	96
79	112		39	96
78	112		38	95
77	111		37	95
76	111		36	95
75	110		35	94
74	110		34	94
73	109		33	93
72	109		32	93
71	108		31	93
70	108		30	92
69	107		29	92
68	107		28	91
67	107		27	91
66	106		26	90
65	106		25	90
64	105		24	89
63	105		23	89
62	105		22	88
61	104		21	88
60	104		20	87

Special Report: What is Test Anxiety and How to Overcome It?

The very nature of tests caters to some level of anxiety, nervousness, or tension, just as we feel for any important event that occurs in our lives. A little bit of anxiety or nervousness can be a good thing. It helps us with motivation, and makes achievement just that much sweeter. However, too much anxiety can be a problem, especially if it hinders our ability to function and perform.

"Test anxiety," is the term that refers to the emotional reactions that some test-takers experience when faced with a test or exam. Having a fear of testing and exams is based upon a rational fear, since the test-taker's performance can shape the course of an academic career. Nevertheless, experiencing excessive fear of examinations will only interfere with the test-taker's ability to perform and chance to be successful.

There are a large variety of causes that can contribute to the development and sensation of test anxiety. These include, but are not limited to, lack of preparation and worrying about issues surrounding the test.

Lack of Preparation

Lack of preparation can be identified by the following behaviors or situations:

Not scheduling enough time to study, and therefore cramming the night before the test or exam

Managing time poorly, to create the sensation that there is not enough time to do everything

Failing to organize the text information in advance, so that the study material consists of the entire text and not simply the pertinent information

Poor overall studying habits

Worrying, on the other hand, can be related to both the test taker, or many other factors around him/her that will be affected by the results of the test. These include worrying about:

Previous performances on similar exams, or exams in general

How friends and other students are achieving

The negative consequences that will result from a poor grade or failure

There are three primary elements to test anxiety. Physical components, which involve the same typical bodily reactions as those to acute anxiety (to be discussed below).

Emotional factors have to do with fear or panic. Mental or cognitive issues concerning attention spans and memory abilities.

Physical Signals

There are many different symptoms of test anxiety, and these are not limited to mental and emotional strain. Frequently there are a range of physical signals that will let a test taker know that he/she is suffering from test anxiety. These bodily changes can include the following:

Perspiring
Sweaty palms
Wet, trembling hands
Nausea
Dry mouth
A knot in the stomach
Headache
Faintness
Muscle tension
Aching shoulders, back and neck
Rapid heart beat
Feeling too hot/cold

To recognize the sensation of test anxiety, a test-taker should monitor him/herself for the following sensations:

The physical distress symptoms as listed above

Emotional sensitivity, expressing emotional feelings such as the need to cry or laugh too much, or a sensation of anger or helplessness

A decreased ability to think, causing the test-taker to blank out or have racing thoughts that are hard to organize or control.

Though most students will feel some level of anxiety when faced with a test or exam, the majority can cope with that anxiety and maintain it at a manageable level. However, those who cannot are faced with a very real and very serious condition, which can and should be controlled for the immeasurable benefit of this sufferer.

Naturally, these sensations lead to negative results for the testing experience. The most common effects of test anxiety have to do with nervousness and mental blocking.

Nervousness

Nervousness can appear in several different levels:

The test-taker's difficulty, or even inability to read and understand the questions on the test

The difficulty or inability to organize thoughts to a coherent form

The difficulty or inability to recall key words and concepts relating to the testing questions (especially essays)

The receipt of poor grades on a test, though the test material was well known by the test taker

Conversely, a person may also experience mental blocking, which involves:

Blanking out on test questions

Only remembering the correct answers to the questions when the test has already finished.

Fortunately for test anxiety sufferers, beating these feelings, to a large degree, has to do with proper preparation. When a test taker has a feeling of preparedness, then anxiety will be dramatically lessened.

The first step to resolving anxiety issues is to distinguish which of the two types of anxiety are being suffered. If the anxiety is a direct result of a lack of preparation, this should be considered a normal reaction, and the anxiety level (as opposed to the test results) shouldn't be anything to worry about. However, if, when adequately prepared, the test-taker still panics, blanks out, or seems to overreact, this is not a fully rational reaction. While this can be considered normal too, there are many ways to combat and overcome these effects.

Remember that anxiety cannot be entirely eliminated, however, there are ways to minimize it, to make the anxiety easier to manage. Preparation is one of the best ways to minimize test anxiety. Therefore the following techniques are wise in order to best fight off any anxiety that may want to build.

To begin with, try to avoid cramming before a test, whenever it is possible. By trying to memorize an entire term's worth of information in one day, you'll be shocking your system, and not giving yourself a very good chance to absorb the information. This is an easy path to anxiety, so for those who suffer from test anxiety, cramming should not even be considered an option.

Instead of cramming, work throughout the semester to combine all of the material which is presented throughout the semester, and work on it gradually as the course goes by, making sure to master the main concepts first, leaving minor details for a week or so before the test.

To study for the upcoming exam, be sure to pose questions that may be on the examination, to gauge the ability to answer them by integrating the ideas from your texts, notes and lectures, as well as any supplementary readings.

If it is truly impossible to cover all of the information that was covered in that particular term, concentrate on the most important portions, that can be covered very well. Learn these concepts as best as possible, so that when the test comes, a goal can be made to use these concepts as presentations of your knowledge.

In addition to study habits, changes in attitude are critical to beating a struggle with test anxiety. In fact, an improvement of the perspective over the entire test-taking experience can actually help a test taker to enjoy studying and therefore improve the overall experience. Be certain not to overemphasize the significance of the grade - know that the result of the test is neither a reflection of self worth, nor is it a measure of intelligence; one grade will not predict a person's future success.

To improve an overall testing outlook, the following steps should be tried:

Keeping in mind that the most reasonable expectation for taking a test is to expect to try to demonstrate as much of what you know as you possibly can.

Reminding ourselves that a test is only one test; this is not the only one, and there will be others.

The thought of thinking of oneself in an irrational, all-or-nothing term should be avoided at all costs.

A reward should be designated for after the test, so there's something to look forward to. Whether it be going to a movie, going out to eat, or simply visiting friends, schedule it in advance, and do it no matter what result is expected on the exam.

Test-takers should also keep in mind that the basics are some of the most important things, even beyond anti-anxiety techniques and studying. Never neglect the basic social, emotional and biological needs, in order to try to absorb information. In order to best achieve, these three factors must be held as just as important as the studying itself.

Study Steps

Remember the following important steps for studying:

Maintain healthy nutrition and exercise habits. Continue both your recreational activities and social pass times. These both contribute to your physical and emotional well being.

Be certain to get a good amount of sleep, especially the night before the test, because when you're overtired you are not able to perform to the best of your best ability.

Keep the studying pace to a moderate level by taking breaks when they are needed, and varying the work whenever possible, to keep the mind fresh instead of getting bored.

When enough studying has been done that all the material that can be learned has been learned, and the test taker is prepared for the test, stop studying and do something relaxing such as listening to music, watching a movie, or taking a warm bubble bath.

There are also many other techniques to minimize the uneasiness or apprehension that is experienced along with test anxiety before, during, or even after the examination. In fact, there are a great deal of things that can be done to stop anxiety from interfering with lifestyle and performance. Again, remember that anxiety will not be eliminated entirely, and it shouldn't be. Otherwise that "up" feeling for exams would not exist, and most of us depend on that sensation to perform better than usual. However, this anxiety has to be at a level that is manageable.

Of course, as we have just discussed, being prepared for the exam is half the battle right away. Attending all classes, finding out what knowledge will be expected on the exam, and knowing the exam schedules are easy steps to lowering anxiety. Keeping up with work will remove the need to cram, and efficient study habits will eliminate wasted time. Studying should be done in an ideal location for concentration, so that it is simple to become interested in the material and give it complete attention. A method such as SQ3R (Survey, Question, Read, Recite, Review) is a wonderful key to follow to make sure that the study habits are as effective as possible, especially in the case of learning from a textbook. Flashcards are great techniques for memorization. Learning to take good notes will mean that notes will be full of useful information, so that less sifting will need to be done to seek out what is pertinent for studying. Reviewing notes after class and then again on occasion will keep the information fresh in the mind. From notes that have been taken summary sheets and outlines can be made for simpler reviewing.

A study group can also be a very motivational and helpful place to study, as there will be a sharing of ideas, all of the minds can work together, to make sure that everyone understands, and the studying will be made more interesting because it will be a social occasion.

Basically, though, as long as the test-taker remains organized and self confident, with efficient study habits, less time will need to be spent studying, and higher grades will be achieved.

To become self confident, there are many useful steps. The first of these is "self talk." It has been shown through extensive research, that self-talk for students who suffer from test anxiety, should be well monitored, in order to make sure that it contributes to self confidence as opposed to sinking the student. Frequently the self talk of test-anxious students is negative or self-defeating, thinking that everyone else is smarter and faster, that they always mess up, and that if they don't do well, they'll fail the entire course. It is important to decreasing anxiety that awareness is made of self talk. Try writing any negative self thoughts and then disputing them with a positive statement instead. Begin self-encouragement as though it was a friend speaking. Repeat positive statements to help reprogram the mind to believing in successes instead of failures.

Helpful Techniques

Other extremely helpful techniques include:

Self-visualization of doing well and reaching goals

While aiming for an "A" level of understanding, don't try to "overprotect" by setting your expectations lower. This will only convince the mind to stop studying in order to meet the lower expectations.

Don't make comparisons with the results or habits of other students. These are individual factors, and different things work for different people, causing different results.

Strive to become an expert in learning what works well, and what can be done in order to improve. Consider collecting this data in a journal.

Create rewards for after studying instead of doing things before studying that will only turn into avoidance behaviors.

Make a practice of relaxing - by using methods such as progressive relaxation, self-hypnosis, guided imagery, etc - in order to make relaxation an automatic sensation.

Work on creating a state of relaxed concentration so that concentrating will take on the focus of the mind, so that none will be wasted on worrying.

Take good care of the physical self by eating well and getting enough sleep.

Plan in time for exercise and stick to this plan.

Beyond these techniques, there are other methods to be used before, during and after the test that will help the test-taker perform well in addition to overcoming anxiety.

Before the exam comes the academic preparation. This involves establishing a study schedule and beginning at least one week before the actual date of the test. By doing this, the anxiety of not having enough time to study for the test will be automatically eliminated. Moreover, this will make the studying a much more effective experience, ensuring that the learning will be an easier process. This relieves much undue pressure on the test-taker.

Summary sheets, note cards, and flash cards with the main concepts and examples of these main concepts should be prepared in advance of the actual studying time. A topic should never be eliminated from this process. By omitting a topic because it isn't expected to be on the test is only setting up the test-taker for anxiety should it actually appear on the exam. Utilize the course syllabus for laying out the topics that should be studied. Carefully go over the notes that were made in class, paying special attention to any of the issues that the professor took special care to emphasize while lecturing in class. In the textbooks, use the chapter review, or if possible, the chapter tests, to begin your review.

It may even be possible to ask the instructor what information will be covered on the exam, or what the format of the exam will be (for example, multiple choice, essay, free form, true-false). Additionally, see if it is possible to find out how many questions will be on the test. If a review sheet or sample test has been offered by the professor, make good use of it, above anything else, for the preparation for the test. Another great resource for getting to know the examination is reviewing tests from previous semesters. Use these tests to review, and aim to achieve a 100% score on each of the possible topics. With a few exceptions, the goal that you set for yourself is the highest one that you will reach.

Take all of the questions that were assigned as homework, and rework them to any other possible course material. The more problems reworked, the more skill and confidence will form as a result. When forming the solution to a problem, write out each of the steps. Don't simply do head work. By doing as many steps on paper as possible, much clarification and therefore confidence will be formed. Do this with as many homework problems as possible, before checking the answers. By checking the answer after each problem, a reinforcement will exist, that will not be on the exam. Study situations should be as exam-like as possible, to prime the test-taker's system for the experience. By waiting to check the answers at the end, a psychological advantage will be formed, to decrease the stress factor.

Another fantastic reason for not cramming is the avoidance of confusion in concepts, especially when it comes to mathematics. 8-10 hours of study will become one hundred percent more effective if it is spread out over a week or at least several days, instead of doing it all in one sitting. Recognize that the human brain requires time in order to assimilate new material, so frequent breaks and a span of study time over several days will be much more beneficial.

Additionally, don't study right up until the point of the exam. Studying should stop a minimum of one hour before the exam begins. This allows the brain to rest and put things in their proper order. This will also provide the time to become as relaxed as possible when going into the examination room. The test-taker will also have time to eat well and eat sensibly. Know that the brain needs food as much as the rest of the body. With enough food and enough sleep, as well as a relaxed attitude, the body and the mind are primed for success.

Avoid any anxious classmates who are talking about the exam. These students only spread anxiety, and are not worth sharing the anxious sentimentalities.

Before the test also involves creating a positive attitude, so mental preparation should also be a point of concentration. There are many keys to creating a positive attitude. Should fears become rushing in, make a visualization of taking the exam, doing well, and seeing an A written on the paper. Write out a list of affirmations that will bring a feeling of confidence, such as "I am doing well in my English class," "I studied well and know my material," "I enjoy this class." Even if the affirmations aren't believed at first, it sends a positive message to the subconscious which will result in an alteration of the overall belief system, which is the system that creates reality.

If a sensation of panic begins, work with the fear and imagine the very worst! Work through the entire scenario of not passing the test, failing the entire course, and dropping out of school, followed by not getting a job, and pushing a shopping cart through the dark alley where you'll live. This will place things into perspective! Then, practice deep breathing and create a visualization of the opposite situation - achieving an "A" on the exam, passing the entire course, receiving the degree at a graduation ceremony.

On the day of the test, there are many things to be done to ensure the best results, as well as the most calm outlook. The following stages are suggested in order to maximize test-taking potential:

Begin the examination day with a moderate breakfast, and avoid any coffee or beverages with caffeine if the test taker is prone to jitters. Even people who are used to managing caffeine can feel jittery or light-headed when it is taken on a test day.

Attempt to do something that is relaxing before the examination begins. As last minute cramming clouds the mastering of overall concepts, it is better to use this time to create a calming outlook.

Be certain to arrive at the test location well in advance, in order to provide time to select a location that is away from doors, windows and other distractions, as well as giving enough time to relax before the test begins.

Keep away from anxiety generating classmates who will upset the sensation of stability and relaxation that is being attempted before the exam.

Should the waiting period before the exam begins cause anxiety, create a self-distraction by reading a light magazine or something else that is relaxing and simple.

During the exam itself, read the entire exam from beginning to end, and find out how much time should be allotted to each individual problem. Once writing the exam, should more time be taken for a problem, it should be abandoned, in order to begin another problem. If there is time at the end, the unfinished problem can always be returned to and completed.

Read the instructions very carefully - twice - so that unpleasant surprises won't follow during or after the exam has ended.

When writing the exam, pretend that the situation is actually simply the completion of homework within a library, or at home. This will assist in forming a relaxed atmosphere, and will allow the brain extra focus for the complex thinking function.

Begin the exam with all of the questions with which the most confidence is felt. This will build the confidence level regarding the entire exam and will begin a quality momentum. This will also create encouragement for trying the problems where uncertainty resides.

Going with the "gut instinct" is always the way to go when solving a problem. Second guessing should be avoided at all costs. Have confidence in the ability to do well.

For essay questions, create an outline in advance that will keep the mind organized and make certain that all of the points are remembered. For multiple choice, read every answer, even if the correct one has been spotted - a better one may exist.

Continue at a pace that is reasonable and not rushed, in order to be able to work carefully. Provide enough time to go over the answers at the end, to check for small errors that can be corrected.

Should a feeling of panic begin, breathe deeply, and think of the feeling of the body releasing sand through its pores. Visualize a calm, peaceful place, and include all of the sights, sounds and sensations of this image. Continue the deep breathing, and take a few minutes to continue this with closed eyes. When all is well again, return to the test.

If a "blanking" occurs for a certain question, skip it and move on to the next question. There will be time to return to the other question later. Get everything done that can be done, first, to guarantee all the grades that can be compiled, and to build all of the confidence possible. Then return to the weaker questions to build the marks from there.

Remember, one's own reality can be created, so as long as the belief is there, success will follow. And remember: anxiety can happen later, right now, there's an exam to be written!

After the examination is complete, whether there is a feeling for a good grade or a bad grade, don't dwell on the exam, and be certain to follow through on the reward that was promised...and enjoy it! Don't dwell on any mistakes that have been made, as there is nothing that can be done at this point anyway.

Additionally, don't begin to study for the next test right away. Do something relaxing for a while, and let the mind relax and prepare itself to begin absorbing information again.

From the results of the exam - both the grade and the entire experience, be certain to learn from what has gone on. Perfect studying habits and work some more on confidence in order to make the next examination experience even better than the last one.

Learn to avoid places where openings occurred for laziness, procrastination and day dreaming.

Use the time between this exam and the next one to better learn to relax, even learning to relax on cue, so that any anxiety can be controlled during the next exam. Learn how to relax the body. Slouch in your chair if that helps. Tighten and then relax all of the different muscle groups, one group at a time, beginning with the feet and then working all the way up to the neck and face. This will ultimately relax the muscles more than they were to begin with. Learn how to breathe deeply and comfortably, and focus on this breathing going in and out as a relaxing thought. With every exhale, repeat the word "relax."

As common as test anxiety is, it is very possible to overcome it. Make yourself one of the test-takers who overcome this frustrating hindrance.

Special Report: Retaking the Test: What Are Your Chances at Improving Your Score?

After going through the experience of taking a major test, many test takers feel that once is enough. The test usually comes during a period of transition in the test taker's life, and taking the test is only one of a series of important events. With so many distractions and conflicting recommendations, it may be difficult for a test taker to rationally determine whether or not he should retake the test after viewing his scores.

The importance of the test usually only adds to the burden of the retake decision. However, don't be swayed by emotion. There a few simple questions that you can ask yourself to guide you as you try to determine whether a retake would improve your score:

1. What went wrong? Why wasn't your score what you expected?

Can you point to a single factor or problem that you feel caused the low score? Were you sick on test day? Was there an emotional upheaval in your life that caused a distraction? Were you late for the test or not able to use the full time allotment? If you can point to any of these specific, individual problems, then a retake should definitely be considered.

2. Is there enough time to improve?

Many problems that may show up in your score report may take a lot of time for improvement. A deficiency in a particular math skill may require weeks or months of tutoring and studying to improve. If you have enough time to improve an identified weakness, then a retake should definitely be considered.

3. How will additional scores be used? Will a score average, highest score, or most recent score be used?

Different test scores may be handled completely differently. If you've taken the test multiple times, sometimes your highest score is used, sometimes your average score is computed and used, and sometimes your most recent score is used. Make sure you understand what method will be used to evaluate your scores, and use that to help you determine whether a retake should be considered.

4. Are my practice test scores significantly higher than my actual test score?

If you have taken a lot of practice tests and are consistently scoring at a much higher level than your actual test score, then you should consider a retake. However, if you've taken five practice tests and only one of your scores was higher than your actual test score, or if your practice test scores were only slightly higher than your actual test score, then it is unlikely that you will significantly increase your score.

5. Do I need perfect scores or will I be able to live with this score? Will this score still allow me to follow my dreams?

What kind of score is acceptable to you? Is your current score "good enough?" Do you have to have a certain score in order to pursue the future of your dreams? If you won't be happy with your current score, and there's no way that you could live with it, then you should consider a retake. However, don't get your hopes up. If you are looking for significant improvement, that may or may not be possible. But if you won't be happy otherwise, it is at least worth the effort.

Remember that there are other considerations. To achieve your dream, it is likely that your grades may also be taken into account. A great test score is usually not the only thing necessary to succeed. Make sure that you aren't overemphasizing the importance of a high test score.

Furthermore, a retake does not always result in a higher score. Some test takers will score lower on a retake, rather than higher. One study shows that one-fourth of test takers will achieve a significant improvement in test score, while one-sixth of test takers will actually show a decrease. While this shows that most test takers will improve, the majority will only improve their scores a little and a retake may not be worth the test taker's effort.

Finally, if a test is taken only once and is considered in the added context of good grades on the part of a test taker, the person reviewing the grades and scores may be tempted to assume that the test taker just had a bad day while taking the test, and may discount the low test score in favor of the high grades. But if the test is retaken and the scores are approximately the same, then the validity of the low scores are only confirmed. Therefore, a retake could actually hurt a test taker by definitely bracketing a test taker's score ability to a limited range.

Special Report: Additional Bonus Material

Due to our efforts to try to keep this book to a manageable length, we've created a link that will give you access to all of your additional bonus material.

Please visit http://www.mometrix.com/bonus948/dsstprstat to access the information.